COAL CULTURES

PHOTOGRAPHY, PLACE, ENVIRONMENT

Series Editors: Liz Wells

Photography, Place, Environment publishes original scholarship and critical thinking exploring ways in which photography contributes to, or challenges, narratives relating to geography, environment, landscape and place, historically and now.

International in scope, and innovatory in placing imagery as both the object and the method of enquiry, the series includes single-authored and edited volumes by new scholars as well as established names in the field. By critiquing relationships between land, aesthetics, culture and photography, the books in this series also foster debates on photographic methodologies, theory and practices.

COAL CULTURES

Picturing Mining Landscapes
and Communities

Derrick Price

Routledge
Taylor & Francis Group

LONDON AND NEW YORK

First published 2019 by Bloomsbury Academic

Published 2020 by Routledge
2 Park Square, Milton Park, Abingdon, Oxon OX14 4RN
605 Third Avenue, New York, NY 10017

Routledge is an imprint of the Taylor & Francis Group, an informa business

First published in Great Britain 2019

First issued in paperback 2021

Cover design: Maria Rajka
Cover image: Naoya Hatakeyama, "Terrils #16015 (Dourges, Évin-Malmaison)" 2010.
© Naoya Hatakeyama / Courtesy of Taka Ishii Gallery

A catalogue record for this book is available from the British Library.

Library of Congress Cataloging-in-Publication Data

Names: Price, Derrick, author.
Title: Coal cultures : picturing mining landscapes and communities /
Derrick Price. Description: London ; New York : Bloomsbury Visual Arts,
an imprint of Bloomsbury Publishing, Plc, 2018.
Identifiers: LCCN 2018030857 | ISBN 9781350037830 (HB : alk. paper) |
ISBN 9781350037847 (epub) | ISBN 9781350037854 (ePDF)
Subjects: LCSH: Coal mines and mining–Pictorial works. | Coal mines and
mining–Environmental aspects. | Mining camps–History–Sources. |
Community life. Classification: LCC TN801 .P75 2018 | DDC 622/.334–dc23
LC record available at https://lccn.loc.gov/2018030857

Typeset by Integra Software Services Pvt. Ltd.

ISBN 13: 978-1-350-03783-0 (hbk)
ISBN 13: 978-0-367-71646-2 (pbk)

CONTENTS

LIST OF FIGURES

ACKNOWLEDGEMENTS

am indebted to Liz Wells for making this book possible. Her incisive reading of the text and her critical responses to it were very valuable. More than this, I need to thank her for the fact that our discussions and debates over the years have helped to shape my view of photography and its place in contemporary culture. The task of sourcing photographs was made pleasurable by working with Sophie Tann of Bloomsbury Academic. She remained positive and constantly helpful even when I was despondent and finding the whole thing a tedious chore. Paul Cabuts was an insightful reader of the manuscript and I really appreciated and benefited from both his encouragement and his criticisms. Richard Dyer very kindly gave me notes he had made on mining in the context of the miners' strike of 1985. Marc Arkless of Ffotogallery patiently guided me through the archive of The Valleys Project, and I am very grateful for his help. My greatest debt is, as always, to my wife, Helen Taylor. Over the years she has driven me through the valleys of South Wales, strapped on a hard hat to travel through heritage mines, and responded to countless stories of the industrial past. Her support for this project was crucial to me, as are the long conversations, jokes, rows and exchanges that are at the heart of our life together.

Markov-Grinberg's 1934 portrait of the 'heroic' miner Nikita Alexeevich Izotov.
© The Francis Frith Collection.

INTRODUCTION

In the sixteenth century the metallurgist Georgius Agricola described the nature and processes of mineral mining in a celebrated and very comprehensive treatise. Published as *De Re Metallica*, it became the standard work on mining for hundreds of years. Much of the power of the work derived from the way in which it was illustrated with wood engravings that helped to explain the technical processes (Agricola 1556). From the first moment of the scientific study of mining, then, visual imagery was central to the description and communication of its nature and processes (Agricola 1556). The photographer and critic Allan Sekula has argued that there is a clear line linking Agricola's famous work to subsequent prints, paintings and, later, photographic depictions of mining (Sekula 1983).

Agricola's study was a remarkable work, but coal mining was centuries old by the time he produced it. Indeed, it is often said that, after agriculture, mining was the earliest of human trades. The Chinese were digging coal out of shallow drifts more than 3,000 years before the birth of Christ. The Aztecs burnt it, and the Romans in Britain, as early as the second century AD, began a trade that carried coal from Newcastle to London. The seventeenth century saw the development of key technical innovations and a hundred years later deep coal mines were established in a number of countries. Long before this, though, coal was important to very early societies. Richard Martin asserts that one of the reasons for China's supremacy in the artistic and technological realms was because their use of coal provided plentiful and cheap energy. He adds that 'Many of the achievements of the later Han dynasty (206 BC to 220 AD) – elaborate lacquerware, exquisite bronze work, the perfection of the papermaking process, and the development of the wind-powered bellows – were made possible by this energy surplus' (Martin 2015: 184).

But it is in the modern world that coal began to take centre stage, replacing timber as the primary source of energy. Indeed, it might be claimed that the mineral brought modernity about. In her influential book on coal, Barbara Freese puts it this way:

FIGURE 1 This woodcut, which shows a mine-pumping device, appeared in Georgius Agricola's *De re metallica libri xii* in 1556.

The industrial age emerged literally in a haze of coal smoke, and in that smoke we can read much of the history of the modern world. And because coal's impact is far from over, we can also catch a disturbing glimpse of our future. (Freese 2003: 2)

The study of coal, then, is a way of exploring the past and predicting something of the future. For coal is a global commodity, and there is no part of the world that remains uninfluenced by it. Even if you could live in a country that neither mined nor burnt coal, global warming and the deleterious atmospheric consequences of burning coal would sooner or later affect you.

The centrality of coal to the making of industrial society means that it has been studied by people in many different disciplines, but our sense that we know something about mines, miners and mining communities comes largely from visual material, especially from photographs.

FIGURE 2 Lewis Hine, Trapper Boy, 1908. Trapper boys waited in the darkness of the mine to open and close doors as wagons passed through. The socially concerned photographer Lewis Hine made this portrait at the Turkey Knob Mine, Macdonald, West Virginia, as part of a long project on child labour.

Coal and photography

This is a book about coal, but also about photography. From its inception photography was concerned to reveal the hidden and dark places of the world. Great cities were trawled for images of the lives of the poor and the dispossessed. Indigenous peoples around the world were captured on camera, as were the noble ruins of vanished civilizations and the minutiae of natural history. But, for early photographers, coal mines were a particular challenge. They were socially and geographically remote from the centres of power, and it was difficult to photograph the work itself, for that took place under the earth in dark caverns that often contained dangerously flammable gases. Nevertheless, there are many early pictures of coal mines, miners, coal communities, and underground working. These were not the first images of mining, for from the eighteenth century there were numerous paintings, engravings and drawings of mine buildings, although they were often pictured from an aesthetic style that stressed their harmony with the surrounding rural scene, rather than the rupture that they made with it. Photography broke this sense that mining was an activity that might easily co-exist with agriculture and rural pursuits.

Coal is a very particular commodity, but it shares many features with other kinds of mineral mining. Gold, copper, nickel and zinc mining have all been extensively recorded, described and photographed. Some fascinating photographic studies have been made of these metals. The gold mines of South Africa are the subject of David Goldblatt's book *On the Mines*. In her introductory essay to this book Nadine Gordimer observes:

> The Witwatersrand created its own landscape out of waste and water brought from the underground in the process of deep-mining, and created its own style of living, inevitably following the social pattern of the colonial era of which it was a phenomenon, but driven by imperatives even deeper than the historical one. The social pattern was, literally and figuratively, on the surface; the human imperative, like the economic one, came from what went on below ground. (Goldblatt and Gordimer 1973: 19)

This connection between the materiality of mining and the underground sources, in both a material and a psychic way, of its power and influence, could easily have been written about coal. Goldblatt's photographs show working miners, but also the artefacts that surrounded them, as well as buildings such as cottages and shops. He also includes a picture of the grand building that housed the company that once financed the mines. If mines are hard to access and comprehend, the patterns of ownership and finance that underpin them are equally opaque and often difficult to understand. The chain of power and control between the mines and the businesses that fund them is the central theme of photographer and critic Allan Sekula's *Geography Lesson: Canadian Notes*, in which he photographed the largest nickel-mining town in the world – Sudbury in Canada. He directly linked these to pictures of the headquarters of the Bank of Canada, whose capital financed the mine (Sekula 1986).

Most books of mining photographs are collections of industrial locales and machinery. Here there are innumerable pithead wheels, cages, trams full of coal, washing plants, coke ovens and slag heaps. Photographs of miners themselves are often of workers in an industrial setting. They have usually just come to the surface, so that they are still black with coal dust and wearing a helmet with a cap lamp. Other scenes show massed men voting for a strike or celebrating some notable event. These photographs are the stuff of academic history. They have often been seen as evidential and can be used to support the veracity of a written account. Forays from documentary photographers were undertaken in order to get beneath the formal surface of things and bring to light the life of the pits and the mining communities.

Miners were significant subjects of American, European and British documentary photography in the 1920s and 1930s, both as proud, indispensable workers and as exemplary figures of unemployment and economic depression. There are countless images of them descending in cages, screening coal on the surface of the pit, or emerging black faced and blinking in the light on their return from the depth of the mine. Underground they are seen hacking at a coal seam with a pick, leading horses or driving machinery. Mining communities were also recorded. Here images of miners bathing in a tin tub before a glowing coal fire are quite common, as are photos of women doing heavy domestic work or cooking meals in a crowded room. Photojournalists were interested in news stories, so tended to concentrate on disasters and strikes. Of course, most pit disasters take place deep underground, so the photographs tend to be of people anxiously waiting at the pithead for news. In addition, a number of major photographers have taken portraits of miners. They are pictured, then, both as distinctive individuals and as social types, so that even a sensitively made portrait may be titled simply, *A Miner*. In the chapters that follow I shall consider various genres of photography and the way in which they have been deployed in picturing these workers. The role of women as miners and as miners' wives, and in embodying the distinctive culture of mining villages, will be an important theme of the book.

Commodities

Many of the products and services that we take for granted, from the food we eat and the clothes we wear to the fuel that creates our power supply, are imported from other countries and have been traded on global markets. Studying the history of the production of these commodities often takes us to formerly colonized nations whose economies have been shaped precisely in order to ensure a steady supply of some vital product. In the service of this people have been enslaved, indentured or exploited. Land has been appropriated and despoiled and particular cultures created. The study of commodities, then, is not simply about economics; it inevitably raises political, cultural and spatial issues. In recent years there has been an upsurge of interest in the history of commodities. Research projects trace the demand in the West for goods such as sugar, cotton, tobacco and tea to the development of slavery and patterns of colonial domination. At the same time there is great interest in popular histories of such things as, salt, potatoes, cod, coffee and coal.

Replacing the burning of trees with an underground source of power was a key moment in human history. It allowed the capitalists of the day to store energy and to use it in different places at appropriate times. This crucial ability to have a regular power supply made modern industrial life possible. In consequence, the future of coal stocks was widely debated and, in the nineteenth century, there was considerable disquiet over the possibility that coal supplies would run out, just as, in Britain, timber had disappeared from the once extensive forests.

In 1866, at a time when Britain was the world's leading producer of coal, Stanley Jevons wrote what was to become a very influential book. In it he contemplated a future when the supplies of coal declined. He made it clear that he was not talking about the total loss of coal supplies in Britain, which he saw as 'literally inexhaustible'. Rather, he was concerned with the future cost of coal, as in order to keep up supplies, pits had to be dug deeper and deeper. He was also worried about the comparative advantage that Britain had over every other country being eroded because:

It is impossible that we should long maintain so singular a position; not only must we meet some limit within our own country, but we must witness the coal produce of other countries approximating to our own and ultimately passing it. (Jevons 1866: 2)

In response to this fear the British Geological Survey worked across the empire to identify new supplies of coal. Indeed, one of the key drivers of imperial expansion was the perceived need to find new sources of the mineral, and the mining history of a number of nations can be understood only within the context of their colonial past. The consequences of colonialism still affect present-day societies. A report from War on Want in 2016 looked at the degree to which British companies control key mineral resources in Africa. More than a hundred companies listed on the stock exchange have mining operations in some thirty-seven sub-Saharan African countries and control more than a trillion dollars' worth of Africa's resources. They conclude that 'the UK government has used its power and influence to ensure that British mining companies have access to Africa's raw materials. This was the case during the colonial period and is still the case today' (War on Want 2016).

Every book about coal stresses its importance in the making of the modern world as it created the steam power that drove the Industrial Revolution. Jevons was right when he said that Britain would eventually be overtaken as the world's largest producer of coal, and other societies would also see it as central to their existence:

Coal made America great. Our nation was built on it – literally and figuratively … For better or worse, the acquisition, transport and processing of coal shaped the development of our cities and reshaped the appearance of the land. Coal determined the location of towns. Coal dictated the routes of canals and railroads. Coal influenced the evolution of society. In recognition of its value, lumps of coal came to be called 'black diamonds'. (Myers et al. 2017: 8)

The importance of coal is diminishing as many nations are pledged to stop burning it in order to reduce carbon emissions. But, in many places in the world, the memory of coal, the people who mined it, and the culture it created still resonate in the collective memory. It is the coexistence of this homage to the memory of coal, together with the material reality of existing and vibrant coal industries, that makes mining a fascinating subject of study for the cultural historian.

Mining in Western Europe is entering the realm of history, and an aversion to coal, and scientific awareness of its polluting properties, are now matters of global concern. International agreements are in place designed to reduce carbon emissions. However, despite great efforts to develop other forms of power, coal remains a major source of energy. In 2017 it still produced about 40 per cent of the world's electricity. It constituted 29 per cent of global energy, making it second only to oil which had a 31 per cent share. Far from declining, since the beginning of the twenty-first century coal has been the fastest growing source of energy in the world. China is a major producer of coal but also consumes half of the world's output. The key exporters are Indonesia, Australia, Russia, South Africa, Columbia and the United States. As in previous centuries, then, it is still a major commodity in the global economy, but its despoliation of the landscape and pollution of the air raises powerful critical voices that want to see it abandoned. Not that this is a peculiarly twenty-first-century problem. Writing in the *Illustrated London News* in 1893, the Rev. Harry Jones considered the prevailing state of coal:

Wise men of science are beating their brains in the search for something to deliver us from the tyranny of coal. That fuel has warmed us and cooked our food for ages – Newcastle was famous for it six hundred years ago. In these latter days it has turned the wheels of our machines, and has given us light from gas. It is difficult to realise what coal does in giving heat, illumination and driving power; let alone the beautiful dyes which the chemist has drawn from the refuse of gas-works. After having been a servant, coal has become our master, without whose aid cities would be left in darkness, our meat would be raw, the railways would be useless tracks of rusty iron, the industry of our factories would cease, and our fleets would be no more able to carry merchandise or to fight an enemy at sea. (Jones 1893: 646)

Musing on the tyranny of coal, and following this immediately with a list of the benefits it delivers, is, in many ways, the contemporary position. Most people and governments agree that carbon fuels must be phased out. Most people and governments around the world still rely wholly or partly on coal for their electricity supply, while people without electricity would happily welcome it whatever its source. Begin to unpick the nature of the industry and statistics rain down from governments, coal companies, independent agencies, medical groups and ecological warriors. Some are clarifying, some tendentious, almost all need to be qualified, and treated with great care. Perhaps most significantly, in predicting the future for coal, analysts are always looking to the past, and comparing how things are now with the days when, as they like to put it, coal was king.

This is a book about coal and about the ways in which photography has contributed to our perceptions of miners and mining communities. Given the long history of mining it is clear that it cannot be an account of the global story of coal and its representation through images and literature. I have examined that history for significant events, and I draw my examples and explore cultural works connected with coal mining from all over the world in order to focus on some key places and incidents that are typical of others.

Types of coal

It is important to know that there are various kinds of coal. In detailed studies of mining the history of coalfields can be explored only in terms of the type of coal that was found there. Different types of coal have diverse and often specialist uses and vary considerably in terms of their economic value. Some are cleaner than others, some burn more brightly and some can provide intense heat. Although it is a mineral, coal comes from organic matter, from plants that have sunk into the earth at various levels and for different periods of time. For centuries people have described coal as a 'living' fuel and have claimed to find in burning it a comforting sense of its origins in natural life. The type of vegetation from which it was formed, together with the depth to which it sank and the temperatures and pressures at those depths, helps to determine the kind of the coal that is produced, as does the length of time it took to create. The youngest coal is a compressed peat, a soft, brown, very dirty coal called lignite that is largely used in power generation. Compress lignite and you produce sub-bituminous coal, which is one of the cleanest. Even older is bituminous coal that is black, shiny and constitutes about half of all coal. It is subdivided into 'steam coal' which was very important when trains and ships were dependent on it, and is still widely used in the generation of electricity and in a range of industrial processes. The second is 'coking coal', an extremely hot fuel which was vital to the creation of steel. Finally, about 1 per cent of all coal is anthracite, the most metamorphosed it is a hard variety with a very high carbon content and few impurities. There is, as a consequence of these different types of coal, no single coal economy. For example, the vast reserves of steam coal in the Welsh valleys meant that its prosperity depended on a buoyant shipping industry; so, in many ways, the exigencies of world trade were more important to its economic condition than the overall state of the British economy.

Coal and politics

Today coal is largely used in power stations to create electricity. President Trump vowed to 'bring back coal', but most experts predict that, despite this pledge, coal will be less important within the power mix in the United States in the future as natural gas becomes cheaper. The technique of fracking continues to lower the price of gas, and in 2016, for the first time, more natural gas than coal was used to generate electricity. About 30 per cent of America's electricity is still produced by coal-fired power stations, but many of these are old and the use of coal, there as elsewhere, would seem to be in terminal decline. However, the position is far from clear: in 2017 the output of coal increased in the United States, China and India. President Trump's aim is to make the United States an exporter of coal, even as the domestic demand falls, an exporter that would rival Australia, currently the world's largest distributer of coal to the rest of the world. China takes a different approach and, instead of exporting coal, it is establishing mines in other countries, some of which have never been serious producers of coal. The Shanghai Electric Group has said that it will build coal-based power plants in Egypt, Pakistan and Iran, with a total output of more than 6,000 megawatts (Tabuchi 2017). Yet everyone agrees that using coal carries serious negative consequences. A leading writer on climate change, Tim Flannery, tells us that 'the burning of coal to generate electricity remains the world's largest single source of carbon pollution. According to the International Energy Agency things are going to get worse' (Flannery 2015: 90).

That coal is not good for human health is not news. For hundreds of years coal and pollution have been linked together, and coal was understood to be the cause of dense fogs, choking industrial atmospheres and air that was deleterious to human health. The debates about the importance of the effects of carbon-based fuels on the earth's atmosphere date from the end of the nineteenth century and have been the subject of serious scientific enquiry since the early 1960s. This research took the idea of the human influence on climate change from the local to the global, so that individual governments meeting in international committees agreed that our dependence on coal and oil must end in the interest of the future of the entire world.

This would all seem to be clear enough, but the present position is very complex. On one hand we have political rhetoric and the signing of international climate agreements. On the other an increasing worldwide demand for electricity, a demand that in many countries can only currently be met by burning coal and, in the case of poorer nations, will not be met for many years without oil and coal. So, many countries have drawn up plans to extirpate coal in the coming decades. Cyprus, Luxemburg, Belgium, Malta and the Baltic countries have all abandoned the fuel. France will close all its coal-fired power plants by 2023, Britain by 2025, the Netherlands by 2030, but, at the same time, a significant dependence on coal continues. Germany is an excellent example of this. It has renounced nuclear energy and put in place plans to become entirely dependent on renewable energy, but without naming a date for the end of coal-burning power stations. At the moment it is still deeply dependent on coal and is home to the biggest mining operation on earth. The Garzweiler strip mine wiped out a number of towns in order to get at the earth beneath them. It stretches for 48 square kilometres and will extract more than 1.3 billion tons of lignite coal before it is closed and the ground flooded to create an enormous lake. Meanwhile, India continues to develop its mining operations, as does Indonesia. Australia is still deeply in love with coal, although its long-term use is problematic, while Turkey plans to build seventy-five new coal-fired power plants in the coming decade. Coal India is the largest coal-producing company in the world. Its vision is to 'emerge as a global player in the primary energy sector committed to provide energy security to the country by attaining environmentally & socially sustainable growth through best practices from mine to market'.

The culture of coal

While the whole world is involved in the coal trade, the culture brought about by the mining of coal differs from one place to another. Looking at the importance of miners in the British imagination, Barbara Freese writes:

> Yet, the symbolic importance of British coal miners comes from more than their once dominant numbers. It comes from the unique mixture of awe, sympathy, guilt and fear that these workers have long inspired. It comes from their work in that most mysterious and dangerous of places, the deep underground, and from their distinctive and isolated tight-knit communities. And it comes from a recognition, at least historically, that coal formed the base of the industrial pyramid on which so much of Britain's greatness rested. (Freese 2003: 234)

The 'symbolic importance' was certainly strong in previous decades. At first sight coal may be seen as lacking any kind of romantic qualities, but its very mundane and grimy ordinariness gave it a kind of plebeian allure that contrasted with the ostentatious commodities of other nations. This is well expressed in John Masefield's poem *Cargoes* that conjures up coastal steamers carrying coal, and lends them a particular kind of homely romance. He compares an ancient vessel, the quinquereme, that he declares is carrying apes, peacocks, ivory and sandalwood, and a Spanish galleon laden with diamonds, emeralds, topazes and cinnamon, with a battered and dirty British coaster hauling coal, firewood, pig iron and cheap tin trays. For all its dullness the coaster is grounded in the world of the everyday, rather than in a realm of luxury. We might also note that transporting coal was central to determining the infrastructure and routes of canals, trains, and steamers. For some people, though, coal is not a grimy product that might be symbolized by a battered coastal vessel, but a modern product. It has often been argued that the introduction of mining to agricultural societies brings with it advanced capitalist modes of production leading to an increase in wages and to training for a workforce. Eventually, it would also lead to greater urbanization and improve the transport infrastructure. At the same time, it would precipitate a loss of agricultural land and result in the despoliation of the rural landscape.

Reading about coal

The literature on coal mining is immense. It covers technical manuals, fiction, memoirs, histories, ethnographic and anthropological studies, medical reports, landscape studies, documentaries, community studies, political science, business news, ecological campaigns, trade union history, working class action, oral history, rhymes, poems, ballads and folk songs. There are also many novels with coal mining as a major theme: from D. H. Lawrence's *Sons and Lovers* to the story of mining communities in the Welsh valleys, Richard Llewellyn's 1939 bestseller *How Green Was My Valley* or Upton Sinclair's exploration of life in an American coal camp, *King* Coal (Sinclair: 1919), to more recent fictional works such as the 1992 novel *Night Ride Home* by Vicki Covington or Martin Cruz Smith's *Rose* which he published in 1996. I shall be drawing on fiction throughout this book, but here I want to introduce just one novel which is perhaps the most important fictional account of mining ever to have been written, Émile Zola's *Germinal*.

Published in 1885, it is the progenitor of all coalfield novels. It covers almost every aspect of mining and much of it is still relevant today. Characteristically, Zola had done a great deal of patient research, and *Germinal* touches on many of the strands of coal mining that became important in later novels and documentaries. The hero, Étienne Lantier, is a newcomer to mining and so has to be taken through all the processes of the trade by one of the women hauliers. So the reader goes with him from the surface, via the rudimentary descent, to the pit bottom and the long trudge to the coalface. Through a graphic description of the back-breaking labour involved, he describes the way in which coal is dug out and taken to the surface – how it is screened, washed and transported to towns and cities. We learn about human labour and the use of horses in the twenty-four-hour cycle of a mine. 'The mine never lay idle: night and day human insects were always down there burrowing into the rock six hundred metres beneath the fields of beet' (Zola 1885: 67). We also learn about the way miners work collectively, usually at this time in family groups, and are relatively autonomous. Indeed, Lantier is only reluctantly accepted because he might be depriving one of the girls in the family of a job: 'The policy of excluding women from working below the surface was anathema to the miners, who were worried about their daughters finding a job and didn't much care about questions of hygiene or morality' (Zola 1885: 31).

Zola makes clear the semi-autonomy of the miner and explains the way in which the 'auctioning of contracts' works, as miners bid for a particular stretch of the coal face to work at a specific price. They functioned as little contractors, paid by output, not by hourly wages. These key facts about mining remained important for many years in the future and Zola unpicks the complex modes of supervision and examines the disputes about the basis of pay which are always a feature of mining life.

The book also tells us about the landscape in which the mine is set, as well as family life, sexual relationships and community values. While we have disasters and fêtes, the central theme of the book is the long running strike, the political stance of the main protagonists and the hardship suffered by the strikers.

Manifesta explores coal mining

Not only are there many books about mining, but the number of cultural works inspired by or created about coal is enormous. At the beginning of the twentieth century large quantities of coal were discovered in the little Belgian town of Ghenk. By 1966 the coal was mined out and the town turned its attention to other industries, leaving behind some colliery buildings. In 2012 a building in the former coal mine of Waterschei was chosen as the site of Manifesta, the travelling European biennial of contemporary art and culture. The theme was mining under the title *Manifesta 9: The Deep of the Modern, a Subcyclopaedia*. The organizers explained that they had chosen the site because it is an elegant Art Deco complex set in a barren landscape and observed that 'as a symbol of the history of labour, it is a massive relic of twentieth century architecture, constructed at the centre of what was once the most industrialized part of Europe' (Manifesta 2012: 13).

Inside this stylish industrial building was the work of hundreds of artists, filmmakers and photographers. They brought together Richard Long's *Bolivian Coal Line* of 1982; Claire Fontaine's 2012 piece called *The House of Energetic Culture*; Maximilien Luce's painting of a mine in Charleroi in 1858; Robert Smithson's land reclamation project, *Nonsite, Site Uncertain*; Janet Buckle's undated but recent painting, *Hatfield: A Working Colliery*; Henry Moore's wartime drawings of miners called *At the Coal Face*; Marge Monko's *Nora's Sisters*, a video piece that looked at gender and class identity, 2009; Marcel Duchamp's *1200 Coal Sacks*, of 1938; and many more. What linked contemporary and historical work from pieces of embroidery, documentary films, postmodern video, technical drawings and conceptual pieces was the fact that they were all about coal. From the mining of it, to its uses, its effluence, its symbolic properties and its importance in shaping whole cultures. Superbly researched and exhaustive as it seemed, there were plenty of artists and movements for which, despite the enormous size of the building, space could not be found.

History and memory

In his celebrated study of archives in his essay 'Photography Between Labour and Capital', Allan Sekula asks how photography can represent the voice of authority while simultaneously claiming to be a token of exchange between equal partners. How can it at once support the status quo and be oppositional and encourage resistance to the existing state of things? Key to understanding this is to ask

> How is historical and social memory preserved, transformed, restricted and obliterated by photographs? (Sekula 1983: 193)

This was one of the key questions asked by the curators of Manifesta 9 who linked contemporary works with those produced in the past and critically examined the history of mines and mining. In this book I also draw on the history of mining, but a number of different types of history are present when making sense of the past. In addition to history as an academic pursuit, we may also see a history that is a study of all the ways in which different groups make sense of the past. Official history is transmitted via a range of characteristic institutions from governments, the media, museums and the academy. The story of everyday life recalled by participants in some event: oral history, personal memory and collective memory are all important to an understanding of mining life. The concept of 'popular memory' is far from being a settled category, and there have been many social and political struggles to define and determine the uses of popular memory. What is clear is that stories from the past often have the power to resonate in the present and help to define the future. Sometimes participants in events may also be interviewed as part of an oral history project. Stories, of major events, family crises, or ordinary existence are recorded and archived in libraries, galleries and heritage sites. The cultural transmission of oral histories is, then, part of the context through which events in the past may be understood and assessed. The digitization of material has led to an increased availability of oral histories. One of the early proponents of oral history, Paul Thompson describes the power of this kind of history as being because:

> The historian of working class politics can juxtapose the statements of the government or the trade union headquarters with the voice of the rank and file – both apathetic and militant. There can be no doubt that this should make for a more realistic reconstruction of the past. (Thompson 1988)

This notion that oral accounts of the past might make for an even-handed history is perhaps too optimistic. It is worth asking who is being interviewed and by whom. Just because one is present at an event does not mean that one attended it with a blank mind, ready to peel off an imprint of disinterested reality. The subject was also influenced by the media, by friends and social contacts, and by the notions of common sense that were current at the time. Reacting to the often-held belief that memory and history are in opposite camps as memory is spontaneous and natural while history is the product of objective thought, the historian Raphael Samuel says:

It is the argument of *Theatres of Memory*, as it is of a great deal of contemporary ethnography, that memory, so far from being merely a passive receptacle or storage system, an image bank of the past, is rather an active shaping force; that it is dynamic – what it contrives symptomatically to forget is as important as what it remembers – and that it is dialectically related to historical thought, rather than being some kind of negative other to it. (Samuel 1994: ix)

This is particularly relevant in the many heritage sites that exist around the world to commemorate a lost mining industry. Here the art of forgetting is as frequently practised as that of commemorative remembering.

In the same spirit, Paul A. Cohen has celebrated the power of story, that is the narratives we tell ourselves. He is less interested in providing a definitive account of what happened at important moments in our history, than in what we believe happened. The stories we tell ourselves of the past are not intended to make sure that we have it right, but because we use these tales to deal with our present existence. So things are written into our memory of the past that may not have happened while correct events may be ignored (Cohen 2014). The construction of post-industrial communities in mining areas draws on a selective version of the past to deal with the dislocations and economic blight of the present. Stories about mining are told in many voices and are validated by very different institutions and modes of authority. There is, of course, a technical literature that describes developments in the business of coal getting from the earliest times. This is a history of machines and the harnessing of power to supplement human labour. But this inevitably coincides to some extent with social history that looks at the way in which new technologies, and the managerial forms through which they are introduced, change the earning power and the way of life of coal miners. Because they live in small, close-knit communities, collective memory is particularly powerful. Until the end of coal mining in Britain the role of miners in the general strike of 1926 was still a live and often-debated issue, as were the dark days of the 1930s. The past was rehearsed through a number of exemplary stories that were familiar to most people. Later these stories became the bedrock of oral history projects and were recycled on television programmes and lodged in archives. In a study of photos 'from the coal era' in the Netherlands, Mariëtte Haveman observes the limitation of photography to reproduce the reality of mining life. Having described the harshness of the working conditions, the industrial injuries and illnesses it promoted, and the family life of those who lived in the shadow of the mine, she concluded:

That reality remained largely out of the picture anyway. Photographs cannot encompass that. But we simply have no better sources, except perhaps the memories of direct witnesses, to convince us that this world once really did exist. (Haveman 2002: 15)

Photographs are often accompanied by supportive texts, and witnesses to the work of the past are represented in words and images. In 2016 the African-American artist La Toya Ruby Frazier took up a residency in the historic Belgian mining region of Grand-Hornu. The last pits had closed forty years earlier, but memories of life in the collieries were still potent. Her book gives moving testimony to the way in which the past shapes and influences the present and her fine images give us a powerful sense of the nature of the place (Frazier 2017).

Photographs and personal testimony may give us evidence about the past, but they both need to be treated with care. They have the immediate appearance of authenticity, but, unsupported, both are unreliable. Personal narratives are subject to the vagaries of time and the idiosyncrasies of the narrator, to the partial viewpoint of the personal, or the pressure to conform to group mores and beliefs. A photograph is notoriously mutable and unstable: its preferred meanings having to be teased out through context, the purpose for which it was made (intentionality), and the other texts and discourses with which it is surrounded. It is for this reason that together with photographs and oral histories, I explore the accounts of mines and mining given in novels, memoirs, official reports and movies.

The structure of this book

While each chapter has a definitive theme I am conscious that each of these offers a plethora of material, with examples that might be drawn from many places around the world. This overabundance of material means that I have had to make hard choices about the things I consider and critique. My concern has been to link a discussion of the practice of mining and the physical and social effects of burning coal with questions of representation and cultural formation. Especially important was a consideration of the role of photography in picturing coal mining communities and, over time, in establishing versions of these largely unknown places in the public imagination.

Chapter 1 begins with a specific slag heap, that central signifier of the activity of mining and of the degradation of the landscape that it brings about. It tracks the progress of the land from the beauty of a pastoral society to the grimy world of mining. It looks at the transitional images that were made as industry took over from rural life. In order to have a vocabulary to discuss the aesthetic and psychic importance of mountains I introduce Burke's notions of the sublime and the beautiful.

Of particular interest is the work of two of the people, John Davies and Naoya Hatakeyama, who have photographed coal tips extensively as well as explored the intrusion of industrial life into the rural and suburban.

Finally, the chapter looks at a coal tip that collapsed with terrible consequences and at the way in which the tragedy of Aberfan was pictured and presented to the public around the world.

Chapter 2 looks at the image of the miner as an archetypal proletarian, then moves to examine the role of women in mining and their exclusion from underground work in Britain after a government report that was primarily intended to consider the position of children in mines. This leads to a consideration of women who worked on the surface of coal mines and became quite famous for their masculine dress and intrepid manner. The collection of photos of these 'pit brow lasses' assembled by Munby is very significant here. I then consider the miner as hero. I look at the importance of Stakhanov in Russian history and at the strategies employed by Soviet artists and photographers in order to picture the proletariat. In the United States quite different methods were used, and I explore these by focusing on the most famous archive of working people, the Farm Security Administration Project of the 1930s, before examining documentary work by more recent American and British photographers and filmmakers.

Chapter 3 is concerned with exploring what a coal community, so often defined as 'unique', might be and how we might best classify them. Here I also consider some of the cultural projects undertaken by miners, especially the Welsh Miners' Libraries. One of the features of coal communities is the constant possibility of accident, as disasters at coal mines are fairly common occurrences. Photography was important in communicating news of disasters, and I look briefly at disaster postcards. I then turn to two films that were central to accounts of disaster in mines – *Kameradschaft*, and *The Stars Look Down*, together with a recent Indian novel, *The Sound of Water*. I examine how people emigrated from or towards mining communities and explore the way in which writers, photographers and intellectuals travelled to mining communities. Perhaps the most famous, George Orwell, is discussed in terms of his well-known book, published in 1937, *The Road to Wigan Pier*. The last section of this chapter discusses mining that takes place in marginal sites that have no or little connection with mining villages. I examine 'artisanal' mining, the Free Miners of the Forest of Dean, 'bootleg' mining and Wang Bing's film *Coal Money* about coal trading in Inner Mongolia. Finally, photographer Chris Killip's book *Seacoal* is analysed in some detail.

In **Chapter 4**, I move away from miners and their communities to look at the pollution caused by coal. Its long history is traced, through particular reference to the smog in Manchester and the famous London fog of 1952 that blanked out the city for five days and led to clean air legislation. I look at the way in which artists, photographers, filmmakers and novelists treated fog from Victorian times. Here, the conflation of atmospheric pollution and moral corruption is significant. Coal gas brought about the lighting of cities and allowed them to function for pleasure and commerce at night. At the same time, coal gas was itself a source of the smog that plagued them.

Chapter 5 is about famous strikes conducted by miners. The British miners' strike of 1984/85 still resonates in contemporary culture and this chapter explores the ways in which it was photographed and filmed. Jeremy Deller's re-enactment of the 'battle of Orgreave', Ken Loach's film *Which Side Are You On?* and Craig Oldham's *In Loving Memory of Work* are discussed, together with the documentary photographs and photojournalism of the time. The second strike - perhaps the most famous in U.S. mining history, The Battle of Blair Mountain – is examined in some detail. We also look at *Matewan*, John Sales' 1987 feature film that restored the events of a mining struggle to a public that had long forgotten them.

Chapter 6 Explores post-industrial landscapes and looks at the work of the Valleys Project in South Wales. It then considers strip mining in Appalachia and moves on to examine mountain top removal. This leads to an account of the American technological sublime. I then examine the concept of the Anthropocene and consider photographers whose work delineates the malign influence of human beings on the natural world. The primary figure here is Edward Burtynsky, but the work of other artists is discussed, including that of Zhao Liang and his influential film, *Behemoth*.

Chapter 7 This final chapter addresses the way in which coal mines are now presented as objects of tourist attention in heritage sites. I discuss the ecomuseum movement before considering some 1980s critiques of heritage that have been of great significance. I then examine the role photography plays in supporting heritage sites as well as the way in which a photography of record functions. Heritage is tinged with nostalgia, an idea I discuss especially via its routes as a diagnosed medical condition and its present status. This leads into a consideration of the nostalgia for coal that some people detect in our time. This nostalgia is commodified in mining souvenirs and given potent expression by the collection of mining banners that were once symbols of collectivity and solidarity. Finally, I look at the ruins of abandoned coalmines that have turned into sites of touristic interest.

1 DEGRADATION AND REGENERATION

FIGURE 3 The town of Bargoed, with its famous coal tip, was photographed as part of the Francis Frith photographic survey of Britain *c.* 1955.

A famous coal tip

When I was growing up in the Rhymney Valley, one of five mining valleys that rise from Cardiff into the hills of South Wales, we were told that the slag heap (we called it a tip) at Bargoed was the largest man-made hill in the world. This was a source of considerable pride to us kids, even though plenty of people cast doubt on the claim. Today an online colliery blog still debates its position in the world league table of slag heaps, from which it emerges as the third largest in Europe – probably. The colliery was opened in the early years of the twentieth century. The

Powell Duffryn Steam Coal Company sank the first shaft in 1897, and the first coal was produced in 1901. Finally there were three shafts and so successful was the venture that in 1909 it broke the world record for the production of coal in a single shift, when more than 4,000 tons were mined. It became the largest coal mine in the Rhymney Valley, and the coal tip grew until, at 400 feet, it could rival the best in Europe. Of course, I am talking about the tip in its heyday, for since the 1970s it has been lowered, re-landscaped and, finally, flattened to make way for a regeneration project and a road scheme. The colliery itself and all the surrounding area have been converted into a country park. In its dark prime the tip loomed over the village of Aberbargoed and the town on the opposite side. It was a huge black mound that attracted attention not only from local people, but also from visiting artists and photographers. Its sheer size gave it a certain grandeur, even though it was nothing more than the dark detritus of the mine dumped over the surrounding hills. All day long buckets on cables rolled overhead to throw more and more waste on to the tip until it stretched away for hundreds of yards across the mountain. It seemed like a natural feature that we all took for granted. Besides it the colliery worked, as they all do, twenty-four hours a day, and the air was full of the sounds of clanking coal trams, hooters and the shunting of trains carrying coal to Cardiff and beyond.

When L. S. Lowry painted a view of Bargoed in 1965 he showed the town spread out in the valley, with a road snaking through it and heavy clouds above. There, on the horizon, is the great loaf-shaped mound of the coal tip. To many people it was an eyesore, but others regarded it as a symbol of the economic success of the town. All this was the culmination of a long history. Iron and coal had been at the heart of the South Wales economy from the eighteenth century, but by the middle of the nineteenth iron smelting was far less important than the mining of coal. As well as having the largest deposits of high-quality steam coal in the world, there were abundant supplies of anthracite and coking coal. It was also a place where innovation and experimentation flourished, so that Wales became a repository of technical skills in mining; skills that the miners took with them when they emigrated to Commonwealth countries and the United States.

The Welsh mining valleys

Despite their importance in the global economy, largely from the production of steam coal, the valleys of South Wales were more or less ignored by the outside world. Many people lived all their lives in Cardiff, but never travelled for twenty or thirty miles to visit them. Like many mining communities around the world, they were hidden places that grew up entirely as a result of the demand for coal, and their once-flourishing local economies have never recovered from the long, slow demise of the extractive industries. Long before the coming of industry these valleys were praised by travellers for their quiet loveliness. Low hills, clear streams, native woodlands and acres of rough heath land made them into places of no economic importance, but of great beauty. All this was to disappear, at first slowly and then very quickly as more and more coal was needed in the nineteenth century. A new, brutal industrial landscape emerged, and South Wales became fixed in the popular imagination as a place of dark mines, grime, polluted rivers, and, of course, of slag heaps. Coal tips are particularly potent, and carry such symbolic weight, because they are a kind of degraded mountain, and mountains were central to the romantic imagination.

In a celebrated and highly influential treatise first published in 1757, Edmund Burke set out his observations on the nature of the beautiful and the sublime. The beautiful is smooth, delicate, well formed, and delights our aesthetic sense. The sublime, though, is associated with pain, with danger and a certain kind of delicious terror. These powerful emotions provoke in us feelings of awe and astonishment as well as deep reverence and respect (Burke: 1757).

The picturesque was a British aesthetic movement that provided a set of rules by which nature should be contemplated and depicted. Beauty in nature was not to be found in formal structures and rigidly patterned gardens, but in the production of 'natural' landscapes that prized wildness and irregularity. Houses, gardens and estates were constructed according to the dictates of the picturesque, and eighteenth-century tourism in Britain was deeply influenced by the search for picturesque views and vistas. Many of the tourist postcards sold to this day conform to its rules. The Picturesque Movement, which was initially formulated in terms of the scenery of South Wales, mediated between the ideas of the sublime and the beautiful. It provided a body of rules that allowed natural landscapes to be viewed like a picture, but the term was applied both to landscapes and the depiction of them, for as Anne Bermingham points out, 'If the highest praise for nature was to say that it resembled a painting, the highest praise for a painting was to say that it resembled a painterly nature' (Bermingham 1987: 57). Now hills were objects of aesthetic contemplation and amateur painters descended on mountainous places in search of scenes that could be viewed, appreciated and recorded according to the rules of the picturesque.

Unlike natural hills, slagheaps were black, and nothing grew on them but patches of anaemic, loosely rooted grass. Fissured by underground streams and surface rain, their instability was apparent to walkers, for the ground slid away underfoot as on a scree. On mountains one might climb higher and higher, look down on the valley below, appreciate the sounds of solitude and contemplate the glories of nature. But it is hard to ramble on a slag heap, and, when active they are noisy, dirty, dangerous places that could be the site of only ironic or bitter contemplation. Innumerable writers commented on the coal tip as a symbol of the degradation of mining landscapes. George Orwell, on his tour of the North of England, put it this way:

A slag heap is at best a hideous thing, because it is so planless and functionless. It is something just dumped on the earth, like the emptying of a giant's dustbin.

On the outskirts of the mining towns there are frightful landscapes where your horizon is ringed completely round by jagged grey mountains, and underfoot is mud and ashes and overhead the steel cables where tubs of dirt travel slowly across miles of country. Often the slag-heaps are on fire, and at night you can see the red rivulets of fire winding this way and that, and also the slow-moving blue flames of sulphur, which always seem on the point of expiring and always spring out again. Even when a slag-heap sinks, as it does ultimately, only an evil brown grass grows on it, and it retains its hummocky surface. (Orwell 1937: 95)

Coal tips are by no means as casual and unplanned as Orwell imagined, but long before him writers had responded to the impact of industry on once-rural places. In his novel *Son of Judith*, Joseph Keating described the relation of the valley to the hills above it:

Above, on all sides rose the hills aglow with the flush of summer; the valley stretched away north and south, fresh and delightful, while a summer haze hung restfully over all the hills and the valley. In contrast with this, the black smoke from the tall chimney stacks, the hideous black furnaces over the pit, the black engine-houses, the black coal wagons, with the clash and clank of heavy couplings and crash of shunting, and the shriek of steam whistles, marked the plague spots of colliery enterprise which buried everything within its radius, houses, river and valley, under black dust, and big tip-heaps of pit refuse. (Keating 1900: 184)

This theme of the contrast between two landscapes is one he develops in his autobiography. Keating was born in 1871 to Irish parents in 'what at that time was a charming, pastoral village in South Wales' and, after working for many years in London as a journalist, returned to the village of Mountain Ash in 1910.

The ancient streets were no longer pleasant and picturesque, but grimy with coal dust ... Hundreds of new streets, long and straight and ugly, and terrible hills of pit refuse, filled the fields in which I had played. The Cynon river was nothing but flowing mud. All semblance of its former silvery winding was gone. (Keating 1916: 269)

In part this is simple reportage, for in the forty years of his absence the face of the valleys did change in a most dramatic way. At that time there was a great boom in coal production and a spectacular increase in population. Older ways of life began to change in response to the new modes of production and the development of capitalism. It would be too simple, though, to see Keating's contrast between the romantic and the sordid, the beautiful and the ugly, as a straightforward, objective account. As always in response to mining communities, he is expressing a profoundly emotional reaction, as well as a simple description. But it is worth noticing that Keating does not see these eulogized fields as sites of harsh agricultural toil, but of innocent play. In fact many farm labourers chose to work in the mines, despite the terrible conditions they had to contend with, rather than stay on the land where the work was also long and hard, and those who carried it out were often sunk in poverty and privation. Keating continues the passage quoted above by observing that 'in many parts the mountains and farms themselves were being buried under pit rubbish. Black industrialism would not stop until it had utterly destroyed the old pastoral life' (Keating 1916: 26). This contrast between the romance of the natural hills and the squalid heaps of colliery waste recurs throughout literature. In one of the most famous novels about coal mining, Richard Llewellyn's *How Green Was My Valley*, the landscape is used to carry the moral resonances of the book, and the central signifier of dark evil is the coal tip. The spoil heap in the book grows larger and larger over the years. It destroys the old, rural beauty; scars the hills that are the site of romance and excitement; and, finally, presses down on the protagonist's house:

> Here in this quiet house I sit thinking back to the structure of my life, building again that which has fallen … The slag heap is moving again. I can hear it whispering to itself and as it whispers the walls of this brave little house are girding themselves to withstand the assault. For months, more than I ever thought it would have the courage to withstand, that great mound has borne down upon these walls, this roof. And for those months the great bully has been beaten, for in my father's day men built well for they were craftsmen. Stout beams, honest blocks, good work, and love for the job, all that is in this house. (Llewellyn 1939: 96)

The anthropomorphized coal tip will, sooner or later, overwhelm the personified house. As industrial labour has replaced craft skill so the malign forces of industry will defeat the 'brave little house'. If the occupant stays much longer, he will be crushed to death as, at the end of the book, his father is crushed 'like a beetle' underground while trying to save the colliery from the revolutionary fervour of a mob (Price 1986).

Degraded landscapes

A slag heap is not a single, contained entity. It goes along with other features of a degraded landscape. John Barr points out that:

> Derelict land sours its surroundings. A spoil tip threatens a much larger area than that on which it perches like some vile bird of prey. A series of heaps or holes in an area kills the interstices as well. Dereliction depreciates – in all senses – the value of the land in its vicinity. It helps to create what we have come to call 'twilight' areas. (Barr 1969: 35)

All over the world mining seriously despoiled the landscape within which it took place. In his celebrated book on mining in Appalachia, Harry M. Caudill observes that often tips were given a lively appearance by the fact that they were burning steadily. Nineteenth-century travellers to Wales, too, sometimes found this a lively sight as they came upon one on a gloomy evening. Caudill sees that on the pit surface (known as 'the tipple' in the United States) mounds of waste were very noticeable.

> Near every tipple there grew up mountainous piles of discarded slate and low-grade coal. As the years advanced these refuse heaps reached heights of hundreds of feet and extended for hundreds of yards in length. As they grew higher chemical processes deep within them caused a spontaneous combustion which set fire to the coal. This fire spread, broke out in other places and burned inexorably day and night. (Caudill 1962: 145)

This reminds us that there are many places where a landscape, once thought picturesque, has been woefully damaged by the coal industry. Caudill uses a highly charged language to describe the depredations of the industry:

> Coal has always cursed the land in which it lies. When men begin to wrest it from the earth it leaves a legacy of foul streams, hideous slag heaps and polluted air. It peoples this transformed land with blind and crippled men and with widows and orphans. It is an extractive industry which takes all away and restores nothing. It mars but never beautifies. It corrupts but never purifies. (Caudill 1962: x)

Even in a work as scientific as the US Survey of the Bituminous Coal Industry, 1947, the effects of coal mining are described using adjectives such as 'scented', 'sparkling', 'pristine' and 'green' to describe the state of the old environment:

> America's coal deposits lie in some of the most beautiful parts of the country. Nature's handiwork, however, has been greatly modified by the enterprise of heedless men. Many valleys, once clean and scented with pine, poplar, and hickory are now filled with the belching of locomotives and a floating haze of grime. Streams that once sparkled and hurried are now choked with silt and sewage. Hillsides once pristine and covered with green, now are scarred with gaping holes, waste dumps and raw gashes that serve as roads. (A Medical Survey of the Bituminous Coal Industry, Supplement: 4)

For Caudill coal is a curse that blights the land, marring its purity and instilling new forms of corruption. In mining country there is always a version of the 'old pastoral' that is a yearned for space long after the work, the people and the values that sustained it are recognized as being irrecoverable. Also, it is interesting that accounts of the effects of mining from Britain, from the United States and from

European countries all use the despoliation of natural landscapes as a metaphor for the moral and psychic degradation of its inhabitants. In *How Green Was My Valley*, what is also pressing against the old way of life, as well as the slag heap, is the presence of 'foreigners' who are described as 'half breed Welsh, Irish and English ... the dross of the collieries'. These people live in one district of the village and the only character to be drawn from their number is a child molester and murderer. Incomers to mining areas are often stigmatized, and all kinds of 'foreigners' have been given a hard time. A steady supply of labour was sometimes hard to find and people often worked without legal rights. Sometimes conditions close to slavery existed, or immigrants were shunted off to the pits without much idea of what they could expect. Recruits might come from a voluntary force fleeing political persecution, or have simply moved from the land, lured by the relatively high wages of mining. It is true that coal mining destroyed the old land wherever it took place, but it also brought about new forms of social existence. People were no longer bounded by place or limited by having to conform to settled ways of life. Freed from the burden of life as a labourer on the land, and detached from a family group, they could begin to negotiate the conditions under which they were prepared to work; that is, they could move towards becoming proletarian wage earners.

Picturing mining

From early in their development, industrial sites, including mining scenes, were quite widely recorded. Mining had long fascinated people, and almost the whole three hundred years of British mineral mining is documented in drawings, engravings, prints, paintings and photographs. During the eighteenth century Wales was the place where a style and iconography for picturing industry were established and numerous painters, including John Hassell, George Robertson and George Samuel, recorded proto-industrial enterprises. Commenting on this activity, Douglas Gray writes:

> These artists ... have bequeathed a major legacy of industrial paintings and drawings that are unique in art history. Amongst them all, they managed to depict almost all the processes contributing to the early growth of Welsh enterprise. Their scenes of coal export, coal pits, copper and lead mines, iron works, forges and tilt hammers, fulling mills, water wheels, industrial housing and steam engines, established a kind of prototype for future depictions of the industrial revolution. (Gray 1982: 14)

Several artists produced bodies of work that looked at the development of science and industry alongside images of the countryside. In 1814 George Walker painted a fascinating watercolour that showed a cheerful-looking strolling miner with a pipe in his mouth, while behind him is a coal mine and alongside him a primitive steam locomotive.

FIGURE 4 This engraving was made in 1815 from George Walker's watercolour of a miner and a steam train.

The public too took a keen interest in the growth of industry and manufacture. From the eighteenth century, despite the gradual despoliation of the natural landscape and a continued interest in the picturesque, fascinated tourists visited industrial enterprises. These were affluent travellers who wanted to see the great engineering and manufacturing sites of the day. However, writing about iron works, Klingender observes that after 1760 'Industry had not yet lost its picturesque character. Still surrounded by romantic scenery, the great ironworks, with their smouldering lime kilns and coke ovens, blazing furnaces and noisy forges, had a special attraction for eighteenth-century admirers of the sublime' (Klingender 1947:8). While industry had its devotees who believed it could legitimately be aesthetically represented, coal was often seen as an unworthy subject for art. For example, Turner's painting of 1835, *Keelmen Heaving in Coals by Moonlight,* was regarded at the time as featuring an ignoble subject on which to waste painterly technical skill. Nor are such opinions confined to the nineteenth century. Reviewing The Arts Council's exhibition, *Coal: British Mining in Art 1680–1980,* Peter Fuller describes art and coal as 'the improbable subject of an exhibition'. He critiques coal not only as the subject of art, but as a potential substance out of which art might be made.

No one ever made a great cathedral, or painted a masterpiece in coal. Coal is uniformly black. It lacks visible variety and is filthy to touch. It has no ornamental value. Coal is too flaky and impermanent to fashion. The qualities it possesses of use to us can only be released by destroying it through burning. Coal offers none of the pleasures of sight, smell, touch or texture which can be derived from working with wood, marble, stone, wool, leather or even paper. (Fuller 1985: 200)

Important though coal may be to the working of industry, Fuller argues that few people have enjoyed the task of mining or moving it. There is, in fact, little 'aesthetic element' in the work. The exhibition was a mix of works that were, in one way or another, connected with mining. Some were, indeed, by miners and he references the Ashington Group, that small group of Northumberland miners who met together for more than fifty years from 1934 in order to draw and paint. They had a number of exhibitions of their work and, in 2007, were the subjects of a play by Lee Hall, *Pitmen Painters*. Many of the objects in the exhibition were not conventional works of art. However, Fuller praises Turner's *Keelmen Heaving in Coals by Moonlight*, because he contends that Turner transformed the plebeian scene before him through 'the play of his artistic imagination and the practice of his pictorial skills'. Fuller considers the curator of the Arts Council exhibition, Douglas Gray, to be overly interested in facticity and social documentation. So, he 'seems to believe that that the best examples of "mining art" appeared with the start of deep mining photography' (Fuller 1985: 203).

Naturally, Fuller was also deeply scornful about John Latham's Carberry Bings project, describing the Bings as 'a large coal tip and an eyesore'. This artwork was a project on a derelict land site in West Lothian, Scotland. In 1976 Latham proposed building large sculptures in the shape of books at the top of spoil heaps or 'Bings'. The coal tips would be preserved and the cost of their removal saved. The project was never carried out. A more ambitious project to transform a coal tip into a piece of art was successfully completed. *Northumberland: The Lady of the North* is an enormous piece of land art. Designed by Charles Jencks, it is described as the largest landscape replica of the female body anywhere in the world. Constructed from coal waste, it opened to the public in 2012. It was intended to be a tourist attraction as well as a central feature in the restoration of a derelict mining landscape.

Peter Fuller is contemptuous of the fact that 'Gray tends to regard as "good" only those works which immediately reproduce the appearance of physical or social reality' (Fuller 1985: 202). The attempt to reproduce the appearance of things was one of the ambitions of early mining photographers. The Arts Council's exhibition contained the work of a number of documentary photographers, including Bill Brandt, Edwin Smith, Edith Tudor-Hart and Kurt Hutton. Indeed, documentary photography was to become one of the major genres through which the life and work of miners were pictured. But there are a huge number of technical and industrial photographs of the whole business of mining from machines, shafts, coalfaces, coke ovens and washing plants, to railway trucks, steam ships and canal barges.

Almost from the birth of the medium, photographers were interested in making pictures of mining, although few of its early practitioners thought of themselves as artists. They did not see themselves as part of a tradition that included paintings of early coal mines by John Laporte or Henry Perlee Parker. There is perhaps little comparison between the picturesque depiction of early coal mines nestling in the middle distance in a scenic view of a heath, and the harshly lit photographs and stereographs of the 1880s. But these, however unsubtle, were technical exercises of high quality. They had to face the central problem of how to shed light on a literally pit-black scene. In 1893 Herbert W. Hughes described the technology available to photographers. Chris Howes quotes him as saying:

> Coal mining photography was hedged around with difficulties … One really flared off as much magnesium as possible, and if the result happened to be good it was good, and that was practically the gist of the whole subject. (Howes 1989: 162)

In 1884 the National Museum in Washington, USA, commissioned a set of photographs that would reveal the conditions that existed in a working mine. They chose a suitable pit – one with relatively easy access and good headroom – and, when magnesium flares proved unsuccessful, used a dynamo to produce electric light. Howes notes that the miners were as impressed as the visitors by this illumination, as they had never seen the mine lit up, but had only glimpsed a few yards ahead with their usual lamps. The photographer, George M. Bretz, managed to make a few acceptable prints, although swirling smoke and coal dust added to the difficulties of the exercise, while in deeper mines the intense heat exacerbated the problems they faced. Many of the early experiments required the miners to pose in static positions with a pick raised overhead or a shovel extended into the loose coal. This gives the workers the fixed and static pose of those models that are so frequently to be found in heritage museums. They also established a visual trope of how to depict miners. Even today it is difficult to represent the conditions of a working mine, beset as they are with intense heat, whirling dust, low light levels, smoke and dangerous gases. Nevertheless, from the start of the twentieth century the demand for photographs of miners grew rapidly, not least from the postcard industry. An extraordinary number of cards picturing mining life were published in the Victorian and Edwardian eras. It was not just in Britain that the subject proved to have considerable appeal with the public – mining postcards were produced in significant numbers in France, Belgium, Germany and the United States. While mainland Europe concentrated on monochrome cards, often including a verse of poetry written by a miner, in Britain and America the tinted postcard proved to be most popular. Views of collieries, groups of miners, pit women, and the various stages of the mining operation all featured on cards (Hannavy 2013: 15). The demand for picturesque views meant that few coal tips were shown, although there was a certain interest in coal pickers on spoil heaps, but all aspects of the industry

FIGURE 5 This stereograph is titled *Slate Pickers, Anthracite Coal Mining, Scranton, Pa., USA*. The cheerful lad in the foreground frames the room where back-breaking labour is taking place.

were covered, including coal docks, domestic coal merchants and cooling towers. Some of these photographs were stereographs (two photographs that gave a kind of 3D image when looked at through a special viewer) and were very popular in Victorian times.

The Bechers

Perhaps the most surprising photographers of coal mines were the Bechers, who are major figures in art photography, famous for their interest in repetition, seriality and a rigorous, disciplined approach to the representation of industrial buildings. Born in Germany in the 1930s, Bernd and Hilla Becher devoted themselves for many decades to a photographic project designed to record European industrial constructions that were falling into disuse, or being swept away as heavy industry declined. They began work in the 1950s, but in the decade that followed the speed of demolition meant that they were in a race against time to document these structures. They travelled widely across Europe recording sites in Belgium, Germany, the Netherlands and the UK.

They treated these artefacts as sculptures and created a typology of industrial buildings, which were to be photographed according to a precise set of rules, and recorded in elegant black-and-white prints. Within twenty years most major art galleries and museums were including the Bechers' images of anonymous objects in their collections: objects which when placed next to each other allowed the viewer to see both the similarities and the intricate differences that existed between them. The notion of seriality derived from the study of nineteenth-century taxonomies of plants and animals, but also from the work of artists in the Weimer republic and the German *New Objectivity* movement. Hilla has said the Renger-Patzsch was an influence, as was August Sander, a coal miner–turned-photographer, who set out to show the face of Germany and recorded hundreds of images of people classified according to their trade or profession. But just as these influences connect the Bechers to the past, they deeply influenced the future of the conceptual art movement which often relied on multiple and serial images. Their method was to construct families of objects, in sequences that connect one image to another.

> Their system is based on a rigorous set of procedural rules: a standardized format and ratio of figure to ground; a uniformly level, full frontal view; near-identical flat lighting conditions … a consistent lack of human presence; a consistent use of the restricted chromatic spectrum offered by black and white photography … uniformity in print quality, sizing, framing and presentation; and a shared function for all the structures photographed for a given series. (Stimson 2006: 146)

There is, then, no meaning inherent in a single photograph of, say, a coal silo or a winding tower, but significance is created by the presence of multiple images. Single, beautifully made photographs are the basic blocks that are then arranged into a series. The emotions they often evoke in the viewer are a mixture of curiosity about the nature of these strange objects and a melancholic sense that they are being lost beyond recovery. There are no workers in these pictures and the once-vibrant industrial sites are still. The captions are straightforward labels of object and place, but we have no social or cultural context through which to view them. We don't even know if they still exist. But the Bechers belong to a particular moment in the development of art and of industrial life. Sculptors such as David Smith and Jean Tinguely had used machine parts as integral to their work. The machine aesthetic and the promise of machine culture that was promulgated by Le Corbusier influenced them, as it did other artists. They were also centrally interested in the nature of photography and the cultural politics of image making.

The seemingly objective and scientific character of their project was in part a polemical return to the 'straight' aesthetics and social themes of the 1920s and 1930s. This new aesthetic was exercised in their many images and in the series of buildings at coal mines. Of course, other people had made images of industry with painstaking thoroughness. For example, the American photographer Carleton Watkins made photographs that were aesthetically pleasing, well composed and produced with forensic detail. While the Bechers knew of his work, they had different aims. Nor did they have any interest in industrial archaeology – that is in using the artefacts of industry to recapture, understand and celebrate their original functions.

Coal mines were central to the Bechers' work and, despite the cool rigour of their approach, it would be wrong to believe they had no emotional attachment

FIGURE 6 The Becher's typological study of Winding Towers in Germany, France and Britain made in 1988.

to the things they photographed. Bernd had come from a mining background and felt he understood the industry and the people who worked in it. When the Bechers visited Bargoed in 1973 they photographed a tower and, in keeping with their disciplined approach, paid no attention to the coal tip lowering above it. Their modernist position was apparently very different from the documentary style that predominated in the photography of industrial areas. With the death of the coal mining industry in Britain, however, the melancholy sight of these once-vital buildings, made of concrete that had become stained and scarred with dust, now stripped of any function but rendered into sculptural forms, powerfully evokes the world from which they were abstracted.

Mining landscapes today

The pastoral landscape was never the site of tranquillity that writers such as Keating would have us believe. The Industrial Revolution in Britain had been preceded by an agricultural revolution that with its enclosures of land and use of machinery displaced agricultural workers and turned the landscape of Britain into one of the least wild and most cultivated in the world, but mines were found in marginal places that had not been subject to agricultural improvement. The early painters depicted a changing landscape, one where the noise of hammers and engines began to intrude on the agricultural scene and the earth began to be blackened by the early industries. Even today, though, the longing for an imaginary idyllic rural past is a powerful element of the national culture. An 'immemorial' landscape of ploughed fields bounded by verdant hedges, or small herds of cows grazing at the edge of a blue sea, decorates many a greetings card and table-mat.

Most photographs of coal tips show them in the background to some more significant image. But some people saw them as objects that deserved to be recorded. The French manufacturer Félix Thiollier gave up his ribbon making business in the 1880s to concentrate on art and photography. He made several powerful studies of slag heaps among his images of the French countryside.

Today, several photographers have taken a more systematic approach to the depiction of the detritus of industry. The British photographer John Davies has pictured the penetration of the urban and the industrial into the rural in a series of books.

In his collection *The British Landscape*, (Davies: 2006) Davies recorded scenes from tourist sites with conventionally beautiful hills and lakes next to power stations, urban centres, railway junctions and collieries. He makes large, formal, usually black

FIGURE 7 John Davies' study of the Terril at Haillicourt, France, 2013.

and white, images that are framed with almost mathematical precision and are often taken from an elevated position. They combine a personal view of the landscape with a documentary impulse to create an accurate study of a place at a particular moment. He produces studies that record the marks of time and the influence of people on the environment. Liz Wells describes his work in the following terms:

> In effect he responded to the challenge of photographing the rural without rendering it picturesque through a strategic reversal; his imagery draws attention to the consequences, residues and margins of industrialization and honours the ordinary through pictorial framing of subject matter previously usually excluded. (Wells 2011: 170)

This commitment to photograph 'the margins of industrialization' is very clear in his collection *Shadow: Slag Heaps of Northern Europe* (Davies 2016). Here he systematically photographed the European spoil heaps (*terrils*), but included earlier images of coal heaps from Britain (including one of Bargoed tip), by way of comparison. The *terrils* he records have often been landscaped into curious shapes as they sit among houses and streets on the flat French plains. The mines that gave rise to them have long disappeared, so that they begin to take on the status of ancient burial mounds, or cromlechs, as objects that signify a distant, irrecoverable past. One at Noeux-Les-Mines has been ingeniously designed to incorporate a ski slope, but for the most part they are simply large earth works that signify only that mining was once at the heart of these communities. Often taken from a high vantage point and in a soft, grey light, they vary from solid black peaks to hills that are wooded and grassed over and will one day completely disappear into the landscape. These images, hinting as they do at a lost society, are both intriguing – for they are intrinsically curious objects – and melancholy, as they arouse in us inchoate feelings of loss.

Rather different sensations are evoked by the photographer Naoya Hatakeyama's study of terrils. He is well known for his interest in the uneasy movement from the rural to the urban that he documented in his native Japan and published in a collection called *Lime Works*. His pictures of terrils are in colour and he teases out from them the maximum aesthetic value, showing coal tips dusted with snow, surrounded at night by illuminated houses, or with a flock of birds flying over them. He makes it clear in notes to the collection that his interest is in both collective and personal memory and the narratives that sustain them. Told by the inhabitants of the local villages that he has come to the region thirty years too late, for there are no mines or mining communities left, he notes that:

> Building on the tiny elements which have survived thirty years into this present future, on recollections which are fading fast, on all the photographs which can be collected and on all the words uttered, we must use our imaginations

to extend the memories of others. If we do not do this, a story will be lost and disappear for ever. (Hatakeyama 2011: n.p.n.)

Less rigorously monumental than John Davies' photographs, Hatakeyama gives us an excellent view of the landscape and the traces of the people who inhabited it. His changing viewpoints and occasional introduction of human figures into the scene offers a clear sense of how it is to be there among what is left of the industrial past. Once loathed, coal tips have now become objects of tourists' curiosity. If, for example, you consult TripAdvisor about things to do in the French city of Lens, your second choice will be to visit Les Terrils Jumeaux, two famous conical-shaped slag heaps that dominate the flat plain. Indeed, the whole of the Nord-Pas-de-Calais itself, with its eighty-seven former mining villages and fifty-one slag heaps, has been declared a UNESCO World Heritage Site.

Hatakeyama is interested in treasuring and preserving collective memories, but, in mining areas, there are some malevolent, even inconsolable, memories with which communities have had to come to terms. Most of these are connected to mine disasters, explosions, fires, floods and gassings. Not infrequently, they have involved coal tips.

Aberfan

All over the world badly constructed or poorly sited slag heaps have failed and broken up leading to the destruction of property or the loss of life. Perhaps the most notorious incident of this kind was at Aberfan in 1966, when a coal tip collapsed sending a sea of slurry down into the valley. It engulfed a school and some houses, killing 116 children and 28 adults. But many other slag heap failures in Wales alone preceded this terrible event. For example, at Rhondda Main in 1928 a tip flowed over the stone wall embankment, tossed telegraph poles to the ground and ripped up the railway line for fifty yards. In 1935, at Fforchaman a slide (which took place only seven years after a previous failure) was described as 'a solid black-grey wave of spoil sweeping relentlessly forward … saturated with water … it drove forward slowly a foul-looking oozing mass. It was like lava sliding down from an erupting volcano'. One could list many other examples in South Wales alone and look at Pentre in 1909, Maerdy 1911, Rhondda Main 1928, Abergorchi 1931, Glenrhondda 1943 and, most significantly, Aberfan itself in 1944 and 1963.

Many people in the local community knew about these collapses and were alarmed by the state of the tip in 1966, but their worries were not taken seriously. They were cautious about complaining too vociferously because they thought that if the tip was condemned the colliery might be closed. This was a time when oil was very cheap and British coal was being out-priced by imports from other European countries. Without too much publicity, a steady series of one-by-one pit closures was moving through the valleys of South Wales.

The tribunal that enquired into the tragedy concluded that there were no villains at Aberfan, but in his Introduction to Gaynor Madgwick's moving memoir of her life as a survivor of the disaster, Vincent Kane argued:

Yes there was, there was one big villain. Coal. King Coal to which we all paid grateful homage in Wales for most of the 20th century. It was coal and the determination to keep producing it at all costs which caused the tip slide; it was coal and the desperate fear of losing it which prompted the dereliction of duty before the disaster and the cover ups and half truths which followed. (Madgwick 2016: 10)

If, as the judge at a tribunal said, Aberfan became a name that needed no explanation, we might ask why this tragedy, in a world full of tragedies, should have resonated in so many people's imaginations. Of course, the fact that large numbers of children were the victims, so that an entire generation had been decimated, was particularly poignant. But, in addition to this, the fact that a close-knit community was the site of the tragedy moved people to sympathy, especially as it became clear that this community was set against the might of the government and an apparently indifferent National Coal Board. Dereliction and neglect appeared to be the birthright of the mining villages, and it was in an area already famous for having lived through hardship and economic depression, whose whole history seemed to be one of privation. It was a tragedy brought about not by acts of God or nature, but through indifference and neglect, and it entered Welsh culture like a wound that has still not healed.

There were worldwide expressions of sympathy and support, and photographers and TV news teams from around the globe travelled to the little village. Even today Getty Images offers a huge range of stock photographs of Aberfan, and there are more than a thousand images on Pinterest. The Magnum photographers Ian Berry and David Hurn both recorded the event. Later Berry said:

It was chaos everywhere. You were trying to keep out of the way because it was pretty emotional. If you were there with a camera it meant that you weren't contributing to the rescue … It was just a case of doing what one could, climbing around in the mud. It was pretty messy and I was trying not to get in the way of people working on recovery. (magnumphotos.com)

Certainly the people of Aberfan quickly grew tired of being photographed, but David Hurn points out that:

Miners digging in this crap to get their children out … do not want a photographer around and performing a job that may prove to have not only historical but legal significance. Nobody can ever say it didn't happen or that it wasn't as bad as it was seen to be. (magnumphotos.com)

FIGURE 8 Mel Parry's photograph taken at Aberfan was seen around the world.

This is the classic defence of documentary photography in heart-rending situations-that the photographs bear witness to the real nature of things. Both photographers produced superb and moving images. Our sense of the nature and scale of the disaster was shaped by television, film and the many photographs that were produced, and fifty years on it is the photographs that seem to get to the heart of the event. Curiously, despite the presence of skilled and creative photographers, a trainee photographer working for a local newspaper, *The Merthyr Express*, took the image that has become most associated with Aberfan. Although he became the 1966 British Press Photographer of the Year, Mel Parry said he couldn't remember taking the photo. In it a policeman carries a child in his arms. He is in the middle of the image and is surrounded by people while the collapsed school stretches out behind him. A woman with an expression of tender concern on her face leans in towards the child. A man carrying an injured child is one of the most familiar tropes in the depiction of suffering and the content of this photograph transcends the fact that it may lack the formal qualities of a fine photograph.

Memory and community

In fact, Aberfan has never been completely out of the news in all the years since the catastrophe occurred. All kinds of people want to visit a place that has become synonymous with tragedy, from social scientists and reporters to casual visitors and disaster tourists. The people of the village were not only bereaved, but were regarded as passive victims of suffering by the media. They were seen as icons of anguish rather than as ordinary human beings, and their collective voice was never really heard. Now, every decade, the media rediscovers Aberfan and combines stock footage of the disaster with interviews with survivors and commentators.

Invited to the village on the fortieth anniversary was the distinguished American artist Shimon Attie, a man who had never heard of the place and had no connections with Wales. His work, though, has always been concerned with the way in which particular places are suffused with special kinds of memory that shape their identity and that of their inhabitants. In Aberfan he was to find an ordinary ex-mining village, typical of many others in the Valleys that had been overwhelmed, not just by the loss of its children, but by the constant pressure of the memory of that event. Where the news crews were anxious to remind us of the perverse specialness of the place, Attie began to explore and celebrate its ordinariness, to construct a picture of the community which took account of, but was not dominated by, its past.

The result was a remarkable multi-screen video installation, which he called *The Attraction of Onlookers* (Attie 2008). The work is impressive for its conceptual sophistication and technical skill. This, however, would be unimportant were it not for the fact that it is emotionally moving and gives the viewer an insight into what it is like to be a part of the damaged community. Working over five months, Attie invited the villagers to be filmed in a makeshift studio. They represented themselves as archetypal figures who embody particular jobs, social roles or positions: there was the Miner, of course, but also the Choir Men, the Single Teenage Mother, the Young Biker, The Boxer and so on. Attie provided a typology of all the kinds of people who make up a particular place. This description makes the piece seem like a nineteenth-century photographic project in which people are recruited merely to illustrate the range of trades to be found on the streets of a great city. But Attie filmed his respondents against a dark background and slowly rotated them through 360 degrees so that, although they are appropriately dressed and carry some suitable props, they emerge not as representatives of a type, but as fully rounded human beings. This generated, simultaneously, a particular kind of anonymity and a special sort of intimacy. This is the heart of the project. It is designed to allow the people of Aberfan to escape the nightmare of the weight of the past, to be ordinary and to remind us what a rich mix of people it takes to make a commonplace community.

Landscape after Aberfan

Aberfan was not just a tragedy; it was also the single most important factor in reshaping the landscape of the Welsh mining valleys in the next decades. There were huge rows between the National Coal Board, the government and the local community about the future of the tip. The authorities were reluctant to pay for

FIGURE 9 A drawing of the Rhymney River at Angel Lane from D. Alun Evans' exhibition, *Black Wound*, 2000.

its elimination and demanded that the disaster fund should bear some of the cost (McLean and Johnes 2000). They did, after a great deal of political infighting, set about making it impossible for another Aberfan to take place. In the following years a huge clean-up was initiated, as part of which the mighty tip at Bargoed was reduced to half its size and landscaped into the side of the hill, although not for another forty years was it to be re-developed.

In 1995 a regeneration project was launched to finally get rid of Bargoed's disused pit, and re-landscape the surrounding area, including the tip. The artist D. Alun Evans, who was then living in the district, became fascinated by this process and spent many long hours with a camera and his sketchbooks recording the tearing away not only of a slag heap, but of a symbol of a whole way of life. The many paintings, photographs and drawings that he made were displayed in an exhibition called *Black Wound*, which was first shown in the Newport Museum and Art Gallery in 2000. This millennial show might have looked back at the older culture with nostalgic regret, but Evans' works are detailed, impartial and apparently dispassionate. The documentary impulse is evident in the close depiction of the processes of change: he showed the land ripped open, traversed by concrete paths and the river flowing through newly constructed culverts. But he also pulled back, so that we see the landscape from across the town, the tip still darkening the horizon. Evans did a great deal of research for this work, studying plans, archival material and, perhaps most importantly, tramping miles over and over the ground in order to record it. In an artist's statement to the exhibition he said:

The rise and fall of the industry in the valleys has been recorded in many forms. Archival fact, personal witness accounts in prose and poetry and many other visual forms underpin the memory. But the transportation of the residue, albeit only a few hundred metres onto the valley floor, has a finality which marks the loss of a daily reminder of the initial industrialisation and exploitation of a naturally beautiful valley. (Evans 2000: 26)

Throughout the valleys coal tips were remodelled to fit into the hills around them, although they are, after decades of being exposed to nature, often recognizably different. A curious flattening of all the surfaces, the anaemic light green colour of the grass and, occasionally, some trees mark the new hills as products of a human agency rather than the 'hand of God', but there is general agreement that the `greening of the valleys' has been a success. This is usually described as 'the return of nature' and the average tourist may not notice the difference between a hill and a greened over tip, allowing us to see an integrated, de-historicized landscape that has 'always' been there and carries no traces of industrial activity. Discussing the position in Britain as a whole, Geoff Coyle points out that:

Despite its colossal scale and virtually nationwide extent, coal mining has left remarkably little legacy in the landscape. The waste tips which were once a prominent part of any coal mining town have been generally contoured and sown with grass to ensure their stability after the Aberfan disaster, or removed for roadstone. The colliery headgears have been dismantled for scrap or to avoid the expense of maintaining them in safe condition. Similarly, winding engines and colliery buildings have gone, and the shafts have been plugged with concrete. (Coyle 2010: 121)

One kind of post-industrial landscape, which has been brought about by conscious reshaping of the earth is, then, abolishing the markers of mining. Where once the coal tips and the black rivers spoke of prosperous industry as well as a despoiled rural beauty, now lakes are stocked with coarse fish and attended by well-equipped anglers. This is a country of vantage places, sign-posted walks and picnic tables, with sensitively placed access roads for cars. Of the industry of mining there are few obvious signs. But nostalgia for an industrial past has extended to fights to retain threatened slag heaps. In 2008 CADW, the body then responsible for heritage in Wales, saved a 6-million-tonne coal tip in Rhostyllen, Wrexham, from demolition. They argued that it was a feature that represented industrial Wales. The stable coal tips that still remain are seen as part of the heritage of the region and their communities. In chapter 3, I look at the nature of these coal communities as they have been established, grown up and developed in different countries, regions and settings. First, though, I want to consider the ways in which miners have been depicted.

2 IMAGES OF MINERS

The dominant image of the miner is of a grime-darkened man wearing a hard hat, holding a shovel and standing next to a cage. But for many centuries miners were, and in some countries still are, women and children, some as young as 5. What do these people do? Most illustrations show a man stripped to the waist hacking at a wall of coal with a pick. In fact, throughout history, most of the people who worked in coal mines never directly hewed coal. In deep mines coal has to be dug out, loaded into tubs and taken in a cage to the surface where it is graded, washed, put into railway wagons and delivered to the customers. Meanwhile, underground, even in relatively un-mechanized mines, the roof has to be constantly propped up, the air tested for noxious gases, ventilation created and maintained, and the roadway driven forward to the next seam of coal. Deep mines, then, are large, labour-intensive enterprises, and employ people in many trades. There are, *inter-alia*, drivers (of horses, mules or machinery), masons, electricians, welders, shot-firers, supervisors, managers, clerks, and surface workers in a variety of jobs. The relatively small percentage of people who work at the coalface, whether with pick and shovel or driving machinery, are the archetypal miners who have been represented in paintings and photographs for many years.

Despite the hardship of the work and the awful environment, mining has always been a relatively popular occupation. In 2017 an online Canadian advertisement designed to attract new workers to the coal industry, and illustrated by two young women, claimed that:

> By working in the coal industry you'll get an above average wage and do interesting and exciting work in some of the most beautiful places in Canada. Skiing, mountain biking, fishing and hiking are just some of the outdoor activities to which you can have easy access. Plus there are many opportunities for career development.

Of course, the work being offered here bears little relationship to the traditional toil of the classic, underground collier, and this is a very different prospectus from that which lured many people into the mines in earlier centuries. But, in the nineteenth century, thousands of people left the land, gave up their lives as agricultural workers and headed for the coalfields. We may think of coal villages as fixed and settled places, spatially and culturally remote and separate, but mining was just one among many working class jobs, and collieries attracted workers when times were relatively good and lost them when things turned bad. Over time real communities were created, jobs passed down through the generations and a particular culture, grounded in the communal activity of mining, was created. But the fluctuating demand for coal always meant that there were buoyant times when it was hard to find enough workers. The United States filled the mining vacancies by internal and external migration, taking many people displaced from Europe or seeking a better life in the new world. In Europe a vast internal migration was the main source of colliery workers, while in India and China today the mines still attract many people away from the land. In response to the constant demand for labourers, miners have been coerced into all kinds of contractual relationships: some were literally slaves; many worked in conditions of servitude; others were well-paid free workers, and some were proletarian fighters in working class movements.

Women in mining

In 1838 there were 216,000 men, women and children working in British coal mines. In that year eleven girls aged from 8 to 16, and fifteen boys from 9 to 12, died in an accident at Husker Colliery near Barnsley. This led, as accidents at mines often do, to a great deal of press attention. Queen Victoria ordered a public enquiry into the incident and in 1840 Lord Ashley persuaded Parliament to set up a Royal Commission of Inquiry into child labour in the mines. Hundreds of testimonies from working miners, officials and doctors were recorded, and they spelled out the nature of the work that had been invisible to most people. The report shocked the nation and led directly to the Mines and Collieries Act of 1842. Not only was the reading public made aware of the horrendous conditions of life and work in mining communities, but the report also drew attention to a lack of religious knowledge together with frequent drunkenness and immorality, all the elements that would stimulate lengthy press reports. What's more, the dreadful conditions were graphically illustrated with wood engravings that showed half-naked women and children crawling on hands and knees to haul trams of coal, climbing ladders, pushing heavily laden wagons with their heads, and shovelling coal.

Although the report was into the treatment of children, it aroused considerable interest in women and the work they were doing in the depth of the mines. Not all collieries employed women underground and William Hunter, a mining overman, commented that when they were employed:

Women always did the lifting or heavy part of the work, and neither they nor the children were treated like human beings; nor are they where they are employed. Females submit to work in places where no man, or even lad, could be got to labour in; they work in bad roads, up to their knees in water, in a posture nearly double: they are below till last hour of pregnancy; they have swelled haunches and ankles, and are prematurely brought to the grave, or, what is worse, a lingering existence. (Report on Child Labour 1842: 50)

Children were beaten, burned, crushed, maimed and killed. Like all miners at the time they suffered from deformations of the body, respiratory diseases, rheumatism, enlargement or hypertrophy of the heart, and ruptures. The Report found that they died at an earlier age than agricultural workers and were more likely to suffer in serious accidents. The Report also drew attention to the fact that men were often naked and women were naked down to the waist. It continued:

Now when the nature of this horrible labour is taken into consideration; its extreme severity; its regular duration of from twelve to fourteen hours daily; the damp, heated, and unwholesome atmosphere of a coal-mine, and the tender age and sex of the workers, a picture is presented of deadly physical oppression and systematic slavery, of which I conscientiously believe no one unacquainted with such facts would credit the existence in the British dominions. (Report on Child Labour 1842: 46)

As a result of the Report in 1842, women were barred from working underground. This was an apparently humane response to the conditions that the Report revealed. But Angela John asks the pertinent question: what were they supposed to do next? (John 1984). They had not gone into the mines from a sense of vocation, nor put up with the appalling conditions out of anything but extreme necessity. Now, spared from the toil of the colliery, but facing destitution, it seems likely that many continued to work underground for some years. Large numbers of women also worked on the surface of the pit where they screened coal (i.e. sorted it from slag), pushed wagons and toiled as general labourers. This heavy work was still open to them, and they began to gain a reputation as female freaks or curiosities. They were known in Lancashire as 'pit brow lasses' and the title was later used in other coalfields. In France and Belgium women who worked on the pit surface were known as *trieuses*, and there are many photographs of these 'coal pickers'. They wore long dresses and headscarves and are usually pictured with clean faces and in carefully posed shots. In Britain the women were dressed in tunics, trousers or layered skirts, and they wore heavy boots or clogs. Often grimy with coal dust, they became objects of great interest to respectable Victorians. Hundreds of photographs were taken of them, and they were pictured on cartes de visite and postcards.

The Munby archive

There are several important archives of photographs of pit brow women, but perhaps the most diligent recorder of them was a barrister, minor poet and amateur painter, Arthur J. Munby (1828–1910). His archive consists of more than seven hundred photographs, almost all of female workers. Munby was obsessed with women who made a living by hard work. He wanted to record

FIGURE 10 This image from the Munby archive is captioned: *Female Collier from Rose Bridge Pits, 1869.*

their clothes, their stance and their coarse work-roughened hands. His duties as a clerk to the Ecclesiastical Commission seems to have left him with plenty of time to travel the country and to talk to, or interview, working women. On a tour to South Wales, armed with his sketchbook, he noted that he had been introduced to 'the grand hills and vallies [sic] of the mineral country, and the splendid chaos of the iron works, and above all, the picturesque ways and frank modest charms of the robust and fearless girls who work at those mountain mines' (Hudson 1972: 211). His collection of photographs, now held at Trinity College Library, Cambridge, is usually described as an invaluable resource for historians, and his many images of the pit brow lasses of Wigan have often been reproduced. They gaze unsmilingly into the camera, sometimes with hands on hips and almost invariably holding the tools of their trade – a shovel or a coal riddle. As well as the women who worked in coal mines, Munby frequently went on urban forays where he talked to all kinds of working women. He often took them to the local photographer where they were immortalized in their working clothes. Historians are uncertain about the nature of the pleasure he took from these encounters. This is made more complex by his personal biography. For many years he employed a woman called Hannah Cullwick as a maid and eventually he secretly married her, although they did not live openly as husband and wife. She is the subject of many of his photographs. He once pictured her as a slave and she habitually called him 'massa'. They both left diaries that describe the many psychosexual games they played in which Hannah adopted a number of identities. Sometimes she was at work on grubby or sordid household tasks; at other times she held him in her arms like a baby.

Victorian society was marked by an extreme fear of pollution and degradation, and its dominant ideas about gender reflect this. In brief, men were to be courteous but active, dutiful and ready to defend women who should aspire to the ideals of modesty, chastity, purity and wholesomeness. Little wonder the bold, begrimed pit brow lasses with their male clothing and labourer's stance inspired feelings from repugnance to lust and *nostalgie de la boue*. Munby, himself, appears to have been driven by a complex of emotions that it is impossible to be authoritative about, but whatever the pleasures his voyeurism afforded him, he did produce a remarkable collection of images.

Women working on the pit top were ideal subjects for photographers because they worked outdoors and were easy to find and encounter. The women and children who worked in cotton or paper mills were guarded from any kind of investigator, confined as they were in large, well-supervised, factories. The pit brow women were exotic in their dress and appearance, and they worked at the entrance to a site of greater mystery and degradation, the pit itself. The revelations of the 1842 report led to a moral panic that established a collective notion of what collieries were like, so that even in the 1860s and 1870s, they were still seen as places of unmitigated filth and moral turpitude. This continued throughout the century and was reinforced when Zola's novel *Germinal* was published in Britain in 1894. At that time women still worked at the coal face in France, and his account of the moral squalor of the mines shocked the public on both sides of the channel.

Women, mining and romance

Fictional accounts of mining life often featured working women as characters in a romance. A good example is the first novel of Frances Hodgson Burnett, who later became famous for *The Secret Garden*. *That Lass O'Lowrie's: A Lancashire Story* was a romantic tale of mining life. It tells of a relationship between a wealthy colliery engineer and a penniless pit brow girl. In the opening paragraphs she describes the group to which her heroine belongs:

> Women who wore a dress more than half masculine, and who talked loudly and laughed discordantly, and some of whom, God knows, had faces as hard and brutal as the hardest of their collier brothers and husbands and sweethearts. They had lived among the coal-pits, and had worked early and late at the 'mouth', ever since they had been old enough to take part in the heavy labour. It was not to be wondered at that they had lost all bloom of womanly modesty and gentleness. (Burnett 1893: 9)

She also sees the pit as a direct source of moral as well as physical decline; in accordance with Victorian ideas about the importance of polluted air, she notes that 'they had breathed in the dust and grime of coal, and somehow or another, it seemed to stick to them and reveal itself in their natures as it did in their bold unwashed faces' (Burnett 1893: 10).

One woman stands out from all this for, although dressed like the rest, she had a striking presence and was, as a consequence of her natural gentility, the butt of jokes and badinage from the others. This is the heroine, Joan Lowrie, who goes on to marry her professional engineer and live a middle class life. This was, I need hardly add, an unlikely outcome to the career trajectory of a pit brow lass.

But, looking at photographs of women who worked on the surface of coal mines it is difficult to see the brutal faces so often described. They are sometimes dirty, although others have dressed up for the occasion. They are solemn, somewhat puzzled and unsmiling (for they lived before the Kodak moment), but their tunics and trousers would scarcely shock us today. These are portraits of women who are underfed and weary, but are otherwise unremarkable. Nevertheless, many contemporary accounts record the extraordinary sight they presented to visitors to mines. In a *Bristol Mercury* article, 29 April 1865, a writer observed:

> In this dress, with faces black with dust and smoke, it is difficult, when elevated 50 or 100 yards, to discern the sex to which these objects belong; and a gentleman, who evidently had never witnessed such a sight before, on visiting the town of Tredegar recently, expressed his astonishment at making mountains

on mountains and enquired what animals those were he saw moving about on the top? In the tempest and storm, in rain and in snow, in the sun and heat, exposed to all weathers, women and girls are employed on the tips in South Wales. (Deller 2013: 4)

Even outside the mine women's domestic work was extraordinarily hard. A report of the US Women's Bureau in the early 1920s described the toil of some half a million miners' wives and unmarried daughters who cooked, washed for and took care of a family, as well as boarders and lodgers. Priscilla Long quotes evidence to indicate that the life was very unenviable. She writes:

The lot of the miner's wife was hard. Physically taxed by frequent childbearing, worn down by unremitting labor from dawn to dusk, these women produced for the time, a significant number of daughters who chose to remain single rather than follow in their mothers' footsteps. (Long 1989: 42)

Today, miners' wives may also be miners, but even in places where there are substantial numbers of women at work in the mines, the job is still seen as strange and unfeminine. Suzanne Tallichet records a woman miner in central Appalachia:

I went into a bank and I wanted to cash my check. The woman [working there] said, 'I want you to sign the name of your husband [on the check]' I said, 'Why?

FIGURE 11 A woman gets ready for work at a Pennsylvania mine. The portrait is by Ted Wathen as part of his work for The President's Commission on Coal, 1979. The original is in colour.

It's mine'. It made me laugh. She said, 'Are you a coal miner?' I said, 'Yes, it's mine'. Then they ask you questions like how can you stand to go down inside there. How can you stand to get so dirty? I say it's all in a day's work. I take a cake of soap and wash it off. That's how I can stand it. (Tallichet 2006: 34)

Not that women were readily welcomed into Appalachian coal mines. It took a lawsuit against the coal companies in 1978 to open up mining jobs to them. Tallichet's respondents also complained that they were given 'the brute work' and were frozen out of the higher-paid jobs that involved running heavy machinery. To add to the continuities of stereotype and discrimination, they also said that many miners' wives disliked them and regarded them as 'whores'.

Masculine bonding

But mining was and remains a predominantly male activity, and among men there was often competition to get the highest output and wages. This called for more than hard work, because the system always tried to reduce the rate for the job. In an oral history project one British respondent recalled the very different wages paid for the same work in different collieries and pointed out:

> There were various reasons for this. You made your own price with management for one. And the other was that some pits would do the same job with a man less so they shared his wage between you … The danger in that was the next job you went on, you made a contract again and they tried to cut you down. (Hall 2012: 387)

Even when a miner negotiated a good contact he was still at the mercy of the systems of weighing and quality control. Hall goes on to say:

> So when you went up the pit they counted how many drams were yours and that's how we were paid. Every dram was a ton and when they checked them if you had a lump of slag in it they wouldn't pay you for that ton. So you had to make sure it was all coal. (Hall 2012: 443)

This was a constant complaint of miners for many years, but competition to be a high-wage man went along with the comradeship of working in small groups and the collective spirit that animated miners. This masculine bonding has been discussed in many ways and is at the heart of the collective spirit of mining. Writing in his 1929 essay 'Mining and the Nottingham Countryside', which is an attack on industrialization and a critique of modernism, D. H. Lawrence described the relationship of men underground:

> Under the butty system the miners worked underground as a sort of intimate community, they knew each other practically naked, and with curious close

intimacy, and the darkness and the underground remoteness of the pit 'stall' and the continual presence of danger, made the physical, instinctive, and intuitional contact between men very highly developed, a contact almost as close as touch, very real and very powerful. This physical awareness and intimate *togetherness* was at its strongest down pit. When the men came up into the light, they blinked. They had, in a measure, to change their flow. Nevertheless, they brought with them above ground the curious dark intimacy of the mine, the naked sort of contact, and if I think of my childhood, it is always as if there was a lustrous sort of inner darkness, like the gloss of coal, in which we moved and had our real being. (Lawrence 1929: 135)

This is a classic Lawrentian take on the collective spirit of the miners and his stress on the 'dark intimacy' engendered by the work may be overstated. Lawrence always claimed that his own father loved mining. Despite being the victim of several serious accidents, he was always glad when he could return to work because of the camaraderie he found there. But miners always had a reputation for more than comradely cooperation. A *Report of the Society for Bettering the Condition of the Poor*, 1798, said of colliers that they were 'naturally turbulent, passionate, and rude in manners and character. Their gains are large and uncertain, and their employment is a species of task work, the profit of which can rarely be previously ascertained. This circumstance gives them the wasteful habits of the gamester' (Thompson 1963: 394).

FIGURE 12 Martin Senior and Tom Cook, miners at Hay Royds Colliery, Yorkshire, 2011, photographed by Ian Beesley and included in his collection *The Drift*, Café Royal Books, 2016.

The miner as hero

Miners, then, were usually regarded as tough, lawless and uncouth, but in the USSR in the 1930s they were to be elevated to the status of heroes. There are relatively few records of the lives of the many thousands of people who have worked underground, but in Ukraine, in the Luhansk region, there is a monument to one particular miner – Alexei Stakhanov. He is shown in full stride with his drill slung over his shoulder like a rifle. He was also pictured on the cover of *Time* magazine on 16 December 1935, and displayed on a Russian postage stamp. He received the Order of Lenin and many other awards and after his death his hometown, Kadievka, was renamed after him. His achievement was to have dug some 112 tons of coal in 5 hours and 45 minutes. This was many times his expected quota, but we now know that he did not achieve this extraordinary feat by any act of astonishing strength or diligence. In fact, what he did was to develop, or demonstrate, a new way of undercutting coal so that it could be quickly loaded. This dramatically changed the prevailing method of production. Indeed, some people have argued that he was helped to his record by collective efforts from a number of miners, with the connivance of the ruling Communist Party that needed a boost both to industrial production and to public morale. Whatever the real story may be, it is notable that a miner was chosen as a symbol of proletarian strength and dedication to the state. For, at that time, in the USSR as in other countries, mining was the bedrock on which all other industries rested.

ДА ЗДРАВСТВУЕТ СТАЛИНСКОЕ ПЛЕМЯ ГЕРОЕВ СТАХАНОВЦЕВ!

FIGURE 13 Gustav Klutsis, *Long Live the Stalinist Order of Heroes and Stakhanovites*, 1936. Original in colour.

Hundreds of photographs of Stakhanov were made: he posed with his drill or jackhammer against a wall of coal; he stood modestly in a room crowded with people who are there to honour him. In 1935 he inspired a famous photomontage by Gustav Klutsis. In this work a large figure of Stalin stands before a crowd of smiling people, with Stakhanov in the front row. Behind Stalin's head is a bust of Lenin. The piece is captioned: *Long Live the Stalinist Order of Heroes and Stakhanovites*. In short, he became the central symbol of a projected new world that could be created only with loyalty to the state, dedication to one's tasks and very hard work. This was so inspirational that to carry out a heroic act of Stakhanovite toil became an ambition for many Soviet citizens in numerous trades and industries.

In fact, Stakhanov was not the first miner to exceed his quota and be hailed as a heroic, aspirational figure. That honour has to go to Nikita Izotov, who in 1932 was also singled out by *Pravda* as a superman who vastly exceeded his expected output. Today he is best remembered for a fine portrait of him taken by the Russian photojournalist Mark Markov-Grinberg (see Frontispiece). In 1934 Grinberg had been commissioned to produce *A Day in the Life* of Izotov for the influential magazine *Soviet Photo*. In one photograph of this series Izotov is covered in coal dust with his miner's lamp slung casually around his neck. He tilts his head upwards and looks into the distance, while behind him is the outline of a colliery building. He is, it almost goes without saying, shot from below so that the heroic stance is emphasized and maintained. This is a fine example of one kind of photograph of miners, one that emphasizes the strength and nobility that flows from male power. This was also a time when photography was the

FIGURE 14 Alexei Stakhanov, 1935, from the ITAR-Tass News Agency. Original in colour.

most potent form of image making and photographs were widely reproduced in a number of European countries. The photo series devoted to Izotov was shown in magazines, including the German *AIZ*, the French *Regard* and the Belgian weekly *Tous* (Vasilyeva 2012).

Picturing workers

In the attempt to take the USSR from a backward agricultural nation to a modern, industrial one in a very short time, morale-boosting images of loyal and successful workers were frequently displayed in picture books and magazines. Writing in *U.S. Camera* in 1938, Edward Steichen commented on how Soviet photographers depicted workers in their publications:

> The Soviet is represented as the home of fine, healthy, vigorous boys and girls marching in Moscow, working in factories, in fields or in clubs; well fed, plump, gay children in government nurseries: lineups of tractors or automobiles giving the impression of miraculously synchronized high speed production. (Steichen 1938: 269)

This was an accurate-enough description of the Soviet output in Stalin's time. However, earlier, in the years after the Bolshevik revolution of 1917, there was a ferment of activity in culture and the arts, and a growth of interest in theoretical questions about the nature of realism, the role of the state in cultural affairs, and the most effective way of depicting the proletariat. These ideas were not merely of interest within the USSR, but were debated, taken up and tested throughout Europe and the United States.

Lenin had declared that cinema was the most important art form for the new Soviets, but the country, sunk in economic turmoil and attacked on all sides, could at first scarcely afford the materials and the infrastructure that are needed to support film and photography. But gradually these limitations were overcome and a number of competing photography projects came into existence (Radetsky 2007).

The first initiative was the establishment of amateur 'worker photographers', a project that flourished in the mid-1920s. These people set out to record on film the pattern of life as it was lived in their localities. Initially they were encouraged in this enterprise, as they offered a way in which social change could be communicated to diverse groups of people throughout a huge empire. As with all photography movements of the time, the existence of well-funded and popular magazines was vital to any kind of success. Perhaps the most important of these was the magazine *Soviet Photo* (*Sovetskoe foto*), founded in 1926. By the end of the decade it sold 25,000 copies every issue. It was a monthly, illustrated magazine with the usual array of articles, photographs, letters, editorials and advertisements.

Later, however, under a new name, it became an organ of state propaganda, espoused socialist realism and moved away from the casual recording of personal experience. As part of this state-inspired change, amateur image-makers were to give way to 'proletarian photographers'. Despite the essential designation of 'proletarian' the individual photographers might well be white-collar workers or, indeed, well-trained professional photographers. The amateur groups had been helped by trade unions who saw photography as an important tool for recording social and industrial life. Now a new infrastructure of the press was gradually established, and this included a number of illustrated magazines with large circulations. Photographers became increasingly important. International, as well as local photo agencies, supported photojournalists, including Arkadii Shaikhet and Semyon Fridlyand, who became famous around the world. The importance of photography was recognized by the state, but it was also seen as a vital art form and, perhaps more importantly, as a skill that everyone should aspire to possess. Anatoly Lunacharsky wrote that 'Just as each vanguard comrade should have a watch, so he should be able to master a photographic camera ... Just as there will be general universal literacy in the USSR, likewise there will be photographic literacy in particular' (Wolf 2011: 36). The high cost of equipment and materials militated against this aim, but it had been hoped that the amateurs would go where no professional photographer set foot, and would reveal hidden, but positive, aspects of social life. The nature of work in factories, farms, quarries, building sites and mines would all be revealed and celebrated. However, in the years of the Cultural Revolution (1928–1932) and the First Five Year Plan, full-time photographers began to take over. Photos by 'worker photographers' seen abroad were likely to be by professionals. There were still groups that supported 'worker photographers'; for example, The Society of Friends of Soviet Cinema also had a photography section, but this was secondary to its work in film and the moving image.

The debate about what kind of photography was appropriate to the new society continued. *Soviet Photo* denounced the work of avant-garde artists such as Rodchenko, regarding them as 'formalist', which was a shorthand term for the idea that they were overly influenced by Western ideas and artists, notably Moholy-Nagy and Renger-Patzsch. Rodchenko fought back in the pages of *Novy lef*. This was a debate between two different conceptions of the function of art and photography, rather than a struggle among social groups, although *Soviet Photo* saw itself as representing 'the people' against those intellectual groups who were influenced by foreign ideas. This debate was brought to an end with the establishment in 1931 of the Russian Association of Proletarian Photo Reporters, a body that aimed to use photography as a weapon in the socialist reconstruction of reality.

The FSA and the 1930s

In New York the Film and Photo League was founded in 1930 as part of the Workers International Relief (WIR), an organization that was committed to aiding striking

FIGURE 15 Children of coal miners, Sunbeam Mines, Scott's Run, West Virginia, photographed by Ben Shahn as part of the FSA project in 1935.

workers worldwide. The Photo League adopted the radical aims of the WIR and was influenced by the international worker-photography movements that emerged in the 1920s and 1930s. This was a period of great interest in photojournalism around the world. Among many factors that helped to develop the practice in the United States was the foundation of the famous photo magazine *Life*, in the same year as The Photo League. Although it began as a rigorously documentary movement, The Photo League members gradually became caught up in experimentation. They embraced, too, the spontaneity provided by using the new, light 35mm cameras, and street photography became important to the group. The Photo League is often seen as a group in thrall to New York and entranced with the task of picturing the bustling life of the streets. But its left-wing statements and stance remained with it and, in 1947, at the time of the Cold War, it was declared a 'subversive organization' and was closed down in 1951. Although The Photo League was an urban movement with little connection to mining, it did raise central questions about the most appropriate ways in which workers should be represented and exercised some influence on the progress and nature of realist photography (Klein, M. 2013).

The Farm Security Administration (FSA) was a government agency that was established in 1935 as part of the Roosevelt administration's New Deal, which was a major attempt to revitalize an economy that was sunk in depression. The photography section was a small part of the FSA that lasted for a decade until it was discontinued in the early years of the Second World War. By 1944 an archive of thousands of photographs of life in America had been created.

Documentary photography was becoming established as the major genre through which social problems could be revealed, pictured and put before the public in magazines, newspapers and government reports. *Life* magazine used photojournalism from 1936, and in both the United States and Europe photo magazines were becoming immensely popular. The FSA was quite well funded, well organized and passionately involved in its work. It made a huge contribution to the scope and power of documentary photography. The roll call of the photographers it employed reads now like a Hall of Fame. In addition to Walker Evans there were Dorothea Lange, Russell Lee, Marion Post Wolcott, Arthur Rothstein, Jack Delano, Ben Shahn, Carl Mydans and Gordon Parks. All of them worked under the direction of the passionately engaged social scientist Roy Stryker. He carefully designed projects, and sent his photographers into the field with specific instructions as to the communities they were to photograph and the social problems they were to illuminate. But once out on assignment, miles from any office, they did the job they were sent to do, but then photographed the world around them, contributing to the archive hundreds of images of the quotidian life of the American people.

When we think of FSA photographs now we tend to see dustbowl country peopled by the rural poor and dispossessed. Dust not only blows across the land, but is etched into the skin and clings to doors and window frames. These pictures are often described as 'human documents' and are lauded for showing 'the face of suffering'. In vain will scholars tell us that there are many photographs in the archive of positive development, of successful social projects, of industrial and technological progress. The faces we remember from the archive are watchful, tentative and marked by signs of poverty and deprivation. The land itself is unstable, the earth friable and fissured. Arthur Rothstein made a number of pictures of cattle drives in Montana with healthy cows and men mounted on fine horses. But the most reproduced of his images is that of a cow's skull resting on cracked earth, a potent symbol of the sad state of things. In the cities a wider range of photographs exist, but they naturally concentrated on slums or poor housing and on displaying the faces of the urban poor.

Stryker supported documentary photography because he believed it capable of showing the real nature of things and of giving unequivocal support to the written reports that spelt out the nature of rural poverty and the crisis of the land. Many of the now-famous photographers who worked for the FSA made images of mines, miners and coal country. There are pictures of drift mines, abandoned mines and successful mines. Miners' homes are represented as well as coal camps. We see miners going into mines, coming home from the shift, eating and chatting. There are photos of miners' wives, of children, schools, shanty towns, slagheaps and tipples. There are few images of miners actually at work, although plenty of them when they are unemployed. It goes without saying that of the coal owners and managers there is no trace. An archive often described as showing us 'the human face of suffering' is emotionally moving, but helps little in looking for the causes of human exploitation. It is also important to remember that, for Stryker, the captions to the photographs were very important. Often they spell out an underlying problem that is not visible in the image itself. For example, Arthur

Rothstein labels his shot of a coal mine as 'Coal mine with strip mining dumps in background: A problem in destructive land use, Cherokee County, Kansas'. Another is titled, 'Coal miner who is unemployed because of mechanization of coal mine'. Not all the photographers were quite so prescriptive, but all were required to support their images with explanatory titles; that is, after all, a central requirement of an archive. Documentary photography rarely tried to create meaning unaided by context and text. It functioned within the discursive rules of magazines, official reports and photo books. Later the archive would be raided for film and television programmes that explored Depression Day America.

Of course, many of the people described as 'miners' were former dirt farmers and agricultural labourers who had left the cracked and fruitless earth in order to find work in the collieries. The movement from land to colliery, from agriculture to mine, is a global one and is happening in some countries today, notably in China and India, even though it has slackened or ceased altogether in European countries and the United States. In the 1930s the mining unions were gaining strength and mechanization was being introduced widely across the industry, but in the FSA archive miners are not revealed to be militant proletarians. They are subsumed into a general rubric of the poor and underprivileged. In his work on *The Archaeology of Knowledge*, Michel Foucault argues that an archive is a system of rules that govern the discourses of what may and may not be said. In other words, exclusion is as important as presence in structuring an archive (Foucault 1972). Today, there is a surfeit of archives and collections of photographs of mines, miners and mining history. They support (*inter alia*) people exploring their family history, local history groups and societies, the heritage industry and lovers of industrial archaeology and photography. In these days of complex databases and

FIGURE 16 Marion Post Wolcott, *Coal Miner, His Wife and Two Children*, Berth Hill, West Virginia, September 1938. The caption adds: 'notice the child's legs'.

the opening up of more archives to scrutiny by the public, archivists have become essential key workers who help to shape and structure our sense of the past. The heritage sites that strive to show mining as it once was depend on photographs both for reference and as display items to support other exhibits.

Documenting three tenant families

One of the most celebrated accounts of rural labour, family life and poverty, *Let Us Now Praise Famous Men*, consists of two quite separate sections: a series of photographs by Walker Evans and a long literary account by James Agee. It is not concerned with mining but I need to consider it because of the influence it had on the way in which workers and the poor were to be represented in future, especially in terms of the relationship between photographs and literary texts. Agee and Walker spent some weeks living with three families of Alabama sharecroppers. They had been commissioned to investigate and report on the state of rural work and poverty. The families whose lives they record, the descriptions of the houses they live in and the work they do link the two accounts. Evans' photographs precede the literary narrative. They are clear, spare, aesthetically superb and set in clear opposition to pictorialist photography.

Agee's prose style and literary strategy owes more than a little to modernism, but it is also dense with allusion and extensive description. He eschews the notion that it might be reportage, just as he renounces the conception that we should see it as art, as journalism, or as fiction. In his frequent glosses upon his own method he describes his aspiration as being to reveal the material existence of the families, together with their emotional and expressive life, without artifice, art or, indeed, without employing any symbolic system. This, he appears to believe, might be achieved by photography which he describes as 'next to unassisted and weaponless consciousness, the central instrument of our time' (Agee and Evans 1969: 11). Indeed, he adds that:

> If I could do it, I'd do no writing here at all. It would be photographs; the rest would be fragments of cloth, bits of cotton, lumps of earth, records of speech, pieces of wood and iron, phials of odors, plates of food and of excrement. (Agee and Evans 1969: 12)

Photography, then, has a privileged place in representation, a system that he viewed as almost as transparent as consciousness itself, that takes us, without the laborious constructions and opacities of language, straight to the heart of things, to what Agee calls 'the cruel radiance of what is' (Agee and Evans 1969: 11). He clearly saw Evans' photographs as having this transparency. However, we might want to note them as having a most sophisticated *mise-en-scène*, and observe that they are not only beautifully and artfully posed, but textured, poignant and, unmistakably, the productions of a particular artist; for while none of them are captioned, all of them are now instantly recognizable as being by Walker Evans. Agee, however, is clear that the camera, in its function as a light that illuminates the most subtle of human relationships, is also unsentimental and concerned only with directly producing truth. He describes Walker Evans assembling his big full-plate camera as 'setting up the terrible structure of the tripod crested by the black square heavy head, dangerous as that of a hunchback, of the camera; stooping beneath cloak and cloud of wicked cloth, and twisting buttons; a witchcraft preparing, colder than keenest ice, and incalculably cruel' (Agee 1941: 332).

The idea that the camera is a kind of witchcraft in that it is an unmediated way of revealing the world was not uncommon in the 1930s. 'I am a camera', wrote Christopher Isherwood, in a work of carefully constructed fiction, 'with its shutter open, quite passive, recording, not thinking' (Isherwood 1939: 7). This is, of course, a very different notion of how the camera functions to that of Agee, not least because Isherwood's metaphor refers to a movie rather than a still camera. But the notion that the camera engendered passivity and a kind of silence in commentators, because of its ability automatically to peel its images from the world in front of the lens, was frequently believed at the time.

If, in the USSR, photography was mobilized by the state to drive forward modernity and industrialization, partly by presenting heroic or inspirational images of workers, it was deployed very differently in the West. Here workers were certainly not seen as heroic figures, but as poor, dispossessed, downtrodden and in need of a square deal. I want to look at two examples of documentary photographs of miners in the United States by considering the work of Russell Lee and Lewis Hine, both of whom in different ways exercised considerable influence on documentary photography.

Russell Lee

Born in 1913, Lee grew up in a small town in rural Illinois and started his working life as a chemist. He came from a comfortably off family and had a private income – a fact that gave him the freedom to pick and choose among the jobs offered to him. He was a key member of the FSA and worked for them from 1936 to 1942. In fact, he created the largest body of work for the organization, producing 23,000 images.

> Lee developed a special approach to documentary photography. It was objective, precise, technically perfect. While some photographers might try to define a subject in one picture, Lee would take an encyclopedic view, building up the visual information through hundreds of selected images. (Rothstein 1986: 50)

This approach of not looking for a single definitive photograph that would leave no more to be said, but of building a narrative cumulatively, suited both the documentary ethos of the day, and the rhetorical devices of illustrated magazines that needed photographs that worked well with literary stories. Lee's dominant conception of the position of workers is interestingly revealed by a quote from an historian of the FSA:

> Where Lange's work beseeched the onlooker to have pity on the poor, Lee's pictures seemed to say, 'Look, here is a fellow who is having a hard time, but he is a decent, hard working man and with a little help, he's going to be alright'. (Hurley 1972: 80)

So, we can see the poor in these Depression Days as in need of our pity, or we can incorporate them into the citizenry at large, making them 'one of us', temporarily down and out, but capable of retaking a place in relatively affluent society. It seems that Lee had a knack for talking to people, connecting with them and getting them to relax in front of the camera. He spent much of his time on the road, living in small hotels and travelling on to the places dictated by Stryker. Sometimes these were coal camps, and during his time with the FSA he made over 4,000 photos of miners. In the Second World War he served in the US Air Force. Aged 38 in 1942 he was too old to be drafted, but the documentary film director Pare Lorentz had been recruited to set up a corps of skilled, professional photographers who would become part the Air Transport Command. He convinced Lee that he should join, so he spent his service taking strategically important aerial photographs of airfields.

In 1947 he undertook a major project working for the *Medical Survey of the Bituminous Coal Industry*, which was the first nationwide medical survey of any industry under the auspices of the national US government. At the time (as at many other times in the United States) the coal industry was in a mess. It was very uneven in terms of the pattern of technological innovation; it was beset by problems of overproduction, followed by periods of scarcity. This was one, but by no means the only cause of terrible relations between management and workers that led to a series of strikes being held in 1946. The authors of the report also recognized that many very different geographical areas produced bituminous coal, and each had its own history, patterns of settlement, modes of working and long-held traditions. They hoped to transcend these differences and, indeed, that the recommendations arising from the research might even be generalizable to other industries. So they were anxious not to treat mining as uniquely disadvantaged. In terms of health provision they wrote:

> Any assertions which may have been made that inferior standards are general in the coal industry are disputed by the Survey. Definitely low standards of health are readily apparent in certain places, but not in all areas where coal is mined. Provisions for health range from excellent, on a par with America's most progressive communities, to very poor, their tolerance a disgrace to a nation to which the world looks for pattern and guidance. (A Medical Survey 1947: vii)

FIGURE 17 Russell Lee, *Milong Bond, Tipple Worker, Taking a Bath*. Even where bathing facilities existed they were often not available for surface workers. Lee took this photograph as part of the Medical Survey project in 1946.

Of course, the report recognized that mining was a particularly dangerous and unhealthy occupation and the Survey diligently explored general medical problems as well as hospital facilities and other agencies of advanced medicine. It also considered the housing stock, and the sanitary facilities. In the report, after a placing shot of a pithead with its winding gear, Russell Lee produced more than a hundred photographs. They are designed to give us a visual record of the deficiencies of the coal villages and camps as well as some of the new and vastly improved facilities. One of the intentions of the survey was to compare the best with the worst, so Lee shows us privies contrasted with indoor bathrooms; houses, roads and gardens that are both good and bad. He also illustrates the medical care of all the family. We are shown children being vaccinated, miners having their lungs checked, a woman in a private room in a hospital and an operating room with surgeons gathered around an anaesthetized patient. Then there is a visual record of domestic life, including images of housework, leisure time, church going and mealtimes.

These are good, unpretentious, photographs that cumulatively build up a picture of mining life, which for the most part is much like the lives of other working people at the time. What is interesting about the project from the photographic point of view is that its authors clearly believed that these images would not only validate their written account, but might be more persuasive. They felt that it is easier to look at two houses – one neat and solid and the other in a terrible state of disrepair – and immediately get the message, than it is to describe the differences. This is one of the best projects of those that were founded on the belief that documentary images could be used, often together with written reports or magazine articles, to produce evidence that would, eventually, be employed to ameliorate the lot of the poor.

One fascinating aspect of the report is a supplement written by Allan Sherman entitled *The Coal Miner and His Family*, which eschewed the technical language and statistical analysis of the main report in favour of a freer, more literary style, as when Sherman notes how many people have come to coal country from around the world:

> Englishmen, Welshmen, Scotsmen, miners themselves and sons of miners; Finns, Italians, Greeks, Magyars, and Slavs and others … went to work in the coal fields. A large number of American born, many of them the sons of the early pioneers who crossed the Cumberland Gap and migrated to Ohio, Kentucky, Indiana, Illinois, and other States forsook the plow for the pick. Then, too, many Negroes, in their search for economic freedom sought out or were brought into the coal fields. (A Medical Survey 1947, Supplement: 2)

'From plow to pick' is a neat way of describing rural migration to the coalfields and the supplement uses the well-worn 'day in the life of' a miner and his family to show us the consequences of abandoning the plough. We go through the diurnal round from breakfast to bedtime, following the miner from the time he changes for work, into the mine itself and along to the coal face, then taking the works bus home, dropping in at a store, arriving home and bathing in a tin tub. Meanwhile his wife is tracked as she goes about her domestic tasks: cleaning, lighting fires, cooking and serving meals.

Noting that modern appliances, such as refrigerators, washing machines and vacuum cleaners, are available to many miners' families, Sherman gratuitously adds that 'They permit her only to have more free time in which to do more work, or more time for aimless calls on her neighbors, and more time in which to be appalled by the nothingness of her surroundings' (p. 30).

This existential cry that claims that free time will be frittered away by women who will themselves be appalled by the world they live in is supported by Lee's image of a woman gazing out into the middle distance through a half-opened window, with a bleakly melancholic expression on her face.

But then we go with the children to school and watch them as they are vaccinated before they do their evening chores of collecting water, chopping wood and shopping at the company store. We see the community at church, drinking at soda fountains and just sitting and chatting. Russell Lee is freer here to provide a photo essay that distances us from Allan's often overblown prose, but provides its own visual narrative of the nature of life in the coalfield. The images still closely interact with the text, but are less literal than those of the survey and allow us to gain a good idea of the character of life in the coal camps and towns.

Lewis Hine

Now one of the most famous of American documentary photographers, Lewis Hine had personal experience of all kinds of working class jobs. His mission was always to drag the processes of labour into the light of day

Alan Trachtenberg asserts that 'Hine's work remained in touch with the history of his times, because:

> What he saw, we now take for granted as definitive realities of American life: the look of emigrants arriving at Ellis Island; of tenements and crowded streets in New York's Lower East Side; and most indisputably, of early twentieth-century American working children – thin, ragged, often wan, more often tough, in coal mines and cotton mills, in field and street and home. (Trachtenberg 1977: 119)

A student of sociology and education, Hine took up photography as an educative tool, but went on to be regarded as one of the foremost documentary photographers in the United States. He was influenced by positivist sociology and by the pragmatism of William James, especially his notion of 'lived experience'. He was also influenced by the world of work and had personal experience of a variety of jobs. As a photographer he worked at a time that was particularly conducive to the kind of revelatory photography he espoused – the period from the 1890s to the 1920s that is known in the United States as 'the progressive era', a time, that is, of social reform, political action and an attempt to haul into daylight the corrupt practices of industrial organizations and government. It was at this time that the word 'muckraker' came into existence, to describe the work of journalists and writers who put before the public stories of venal or crooked conduct. Among these was the writer Upton Sinclair, whose novel *King Coal* vividly described life in a coal camp and militant industrial action based on the 1913–1914 Colorado coal strike. Also enlisted under the muckraker title was Jacob Riis, whose accounts of the life of the poor and the striking photographs that accompanied them stirred public sympathy for the lot of immigrants in New York City. Hine almost certainly knew Riis' work and like him he believed that the visual was instantly revelatory and that the public would respond to direct images in a way not possible in literary accounts. For them seeing was believing and in consequence the camera was viewed as a central instrument through which social change might be achieved. But Hine's work is far more complex and subtle than that of Riis, who had little or no interest in the aesthetic function of his images and would never have described himself, as Hine did, as an *interpretive photographer*.

In 1907 Hine worked on *The Pittsburgh Survey*, one of the first and most important surveys of industrial work in the United States. A year later he joined the National Child Labor Committee, and for the next ten years he recorded child labour in mines, cotton mills and factories. Often in danger from guards, he had to use disguise and misrepresentation in order to gain access to these work places. In the spirit of 'the survey' he aimed to pile example upon example, so there are dozens of little girls toiling in cotton mills and small boys working in coal mines. He also recorded young cotton pickers and workers in glass factories. Hine believed that authoritative meaning would grow from the accumulation of evidence, but we are also invited to look at the human face of toil. The dark lines of fatigue are etched on innumerable young faces; the stooped and arthritic shoulders of pit boys working on the breakers recur in many photographs. Meanwhile, industrial methods were changing rapidly. In 1908 Henry Ford introduced the Model T car, which was the first mass-produced, standardized vehicle. It depended on the smooth running of a production line with a high degree of linear division of labour. Ford thus kept the price low, but also paid his workers well so that they could one day perhaps afford to buy his products. The 'scientific management' notions of Frederick Taylor influenced Ford. Taylor broke work down to its component parts and timed each activity with great precision. In this system the worker was no longer even partially autonomous, but became part of the industrial system and was required to carry out a single task repeatedly and in a given time so as to maximize the output of a machine. These moves to a modern factory system coexisted with the direct exploitation of workers, including children in other industries. In 1916 Woodrow Wilson, influenced to a large extent by the campaigning work of the Child Labor Committee, passed legislation prohibiting child labour. But the Act was deemed to be unconstitutional and not until 1938 was a new version of it finally accepted. This difficulty in bringing about social change raises again the question of just how powerful the camera can be in revealing the roots and causes of poverty and exploitation in a way that alters things. Even critics of Hine's photographs acknowledge his determined attempt to transform the world of work. Stuart Franklin, however, has critiqued Hines saying that 'Hine considered himself an artist, but the dullness of many of his compositions is rescued only by the handful of pictures that break his formulaic mould and by his tenacity as a social campaigner.' (Franklin 2016: 59)

Franklin sets Hine's aesthetic and technical skill against his work as a social activist. However, a ferment of work with charities, journals, government, and educators was the world in which Hine thrived. It was the frame that structured his pictures and through which they achieved coherence and emotional depth. This, though, was also a time when the great cities were powerful centres of action and concern. They were at once the sites of wealth, power, influence and excitement, and of poverty, slums and deprivation. They were the locales both of social evils

FIGURE 18 Lewis Hine, *Breaker Boys*, 1911 (123).

and of enlightenment. The huge influx of new immigrants gave American cities a particular character, but in Britain too in the early work of Mayhew, Thomson and Annan, cities are seen as places both of social evil and of potential enlightenment. Social facts, it seemed, could be found at their most stark and vivid in the sprawling, overcrowded and disorganized social space of the city. This makes Hine's work in the mines of particular interest. He moved away from urban life and travelled the country recording breaker boys in Pennsylvania, trapper boys in West Virginia and young miners in Tennessee and Alabama. The most often reproduced of these photographs are the breaker boys.

These were young lads who worked on the tipple and were engaged in the back-breaking labour of picking coal from the slag and dross that also made its way to the surface. Crouched over their work they stooped for fourteen hours a day in the breaker sheds, which were so full of dust it was difficult to breathe. Hines shows them at their task with an overseer standing at the back of the room and carrying a stick. But he also filmed them lined up for the camera with their faces, hands and clothes black with coal dust, or posing for individual portraits. It might be thought that these images were unequivocal evidence of the nature of these lives, but Hine was aware that, despite the power of the visual, he would, like all documentary photographers, be challenged as to the authenticity and generalizability of his work. The conception of 'the survey' helped with this, but to be effective, the images had to be incorporated within other discourses, so they were captioned, accompanied by text, supported by statistics and reinforced by the authoritative accounts of experts. This desire to support documentary work with an abundance of facts, figures and other evidence was employed in a very different way in the British documentary movement.

British 1930s documentary

The film movement in Britain was of great importance in establishing a particular kind of documentary. It was one that spoke to a putative audience that was keen to know more of the nature of its own society. It worked together with other characteristic movements, magazines and organizations from Mass Observation and *Picture Post* to *The Daily Mirror*. In 1933 the GPO Film Unit was established. Headed by John Grierson, who coined the word 'documentary' to describe his version of realist film, it was set up to create sponsored documentary films and was much influenced by Soviet cinema. Its first production was Alberto Cavalcanti's poetically charged 1935 film, *Coal Face*, with music by Benjamin Britten, and verse by W. H. Auden. The film is a mixture of narration and related images and, in the absence of sync sound, the voice-over gives us facts. The opening statement with music behind it booms out the fact that coal mining is the basic industry of Britain. There is what was to become in many films the placing shot of a pithead. Single statements now illustrate what we are seeing on the screen. The narration is strictly factual: at the surface the coal is washed and graded. The principal by-products of coal are coke, gas, tar, dyes, oil, benzol and so on. We see miners walking into work and move into close-ups of the business of cutting coal below ground. The titles of the various jobs in the colliery, banksman, overman, collier, inspector and so on are rhythmically chanted. There is a mélange of sounds with miners saying a few sentences about the day's work. We see an 'electric' coal cutter at work and statistics are reeled off: one in every five miners in Britain injured every year; 40 million tons of coal a year are for household use, 10 million for electricity, 12 million for steam trains, 15 million for shipping, 54 million exported and 85 million for industry. The film goes on to tell us that after work life is bound up with the pit. The life of the village depends on the pit. Then we see the coal when it leaves the pit. Coal train wagons are running to the rhythmic sound of a drumbeat, overlaid with the noise of a steam whistle. Finally the film reminds us again that coal mining is the basic industry of Britain. Despite the statistics, and the footage of miners at work and at home, this is a film that uses music and a poetic rhythm to create a particular sense of strangeness, of distance from the subject itself, the better to drive its single sentence messages home. R. M. Barsam points out that it was an experimental forerunner to the very successful *Night Mail* which used the same mix of poetic script integrated with music, but adds that:

> As an explanation film about the processing and distribution of coal it is strident … Important, however, and unique among these early British films is the strong voice of social protest that comprises the film's message. The chorus of oppressed miners relates its submission to the pits with a tone of bitterness and futility. (Barsam 1974: 50)

Despite their brief, GPO films tended not to be content to be 'explanation' films but used many visual and sound devices to deliver a highly charged reality that owed as much to the rhythms of art as they did to the presentation of statistics.

While the GPO film unit concentrated on the importance of coal to the nation they did not engage with the fact that the industry was in recession and many mining communities were in a desperate economic plight. These were the 'depressed', 'distressed' and 'special' areas that were the subject of so many reports, surveys and sociological studies. Some photojournalists concentrated on images of the once-powerful miner as now down-and-out and on the breadline.

The 1930s produced a number of talented British photojournalists and documentary photographers. Well known today is Bert Hardy, who worked for *Picture Post*, from 1941 and photographed around the world, but is perhaps best remembered for his photos of slum life and images of the bombing of Britain. More politically charged was the work of the Austrian photographer Edith Tudor-Hart. A communist party activist, she had studied at the Bauhaus and she went on to photograph miners in Tyneside and South Wales together with children in the London slums. Her work is direct and socially critical. Commenting on her in the introduction to a collection of her pictures, Christopher Baker and Wolfgang Kos

FIGURE 19 An unemployed coal miner in Wigan pictured in the late 1930s by an unnamed photographer.

say: 'In Britain her grounding in the realist dynamic of continental photography lent her work a special charge; British photographers of the thirties appear rather dilettantish by comparison' (Tudor-Hart 2013: 9).

While there may have been dilettantes among their number, several British documentary photographers were seriously engaged on complex projects at this time. They tended to shun either the simple presentation of social facts or the creation of politically charged photographs. Rather, they strayed on the edges of Surrealism and frequently produced enigmatic images that refuse simple readings. This can be seen in the work of the German-born Bill Brandt, who was very influential in the 1930s and 1940s for both his photojournalism and his documents of British life in a domestic setting. An image of a miner in a tiny tub being washed down by his wife is placed in tendentious juxtaposition with one of a maid in frilly cap and apron drawing a bath. Both women stoop over their task, in the gestural curve of their backs they echo one another. Brandt's work has an unsettling and ambiguous air. These qualities can also be observed in the photographs of Humphrey Spender. Although his time as a photographer was quite short, Spender managed to make pictures for some characteristic British organizations. In the late 1930s he was employed as 'Lensman' on the *Daily Mirror* when he was asked by Tom Harrisson to work for Mass Observation on a project to record the lives of people in Bolton, which they called 'Worktown'. Initially he was shocked by conditions in the North and by the grime and smog of the city, but he produced some extremely interesting photographs (Spender 1978).

FIGURE 20 A woman pegs out her washing in Stephen Spender's photograph taken in Yorkshire in 1936.

It is hard to discuss the work without referring to his background as an upper-class privileged man, not least because in interviews he always spoke about his feelings of unease in the company of workers. He also took to extremes the Mass Observation precept that people should not know they were being studied. He described himself as 'a spy' and was very aware of the social distance between him and his subjects. Although he is usually described as a social documentary photographer, Spender's images are hard to categorize in this way. He saw himself essentially as an artist, had lived in Germany and was more than a little influenced by Surrealism. It is easy to see curious and disjunctive objects in his photographs, but also the curious stillness that pervades the streets of 'Worktown', a busy manufacturing town, in some of his shots. Ian Walker has drawn attention to this and noted the influence of de Chirico (Walker 2007). Spender, too, photographed coal breakers, and in 1934 he travelled to Northumberland and photographed at a mine near Ashington.

The critique of documentary

Documentary film and photography became the major genre through which the lives of the poor and the working class were represented. But the form was frequently criticized. Commenting on the photojournalism and documentary photography of the 1984/1985 miners' strike in Britain, Stuart Hall said:

> I think that what has happened is that the left has got stuck in a certain period, which is why the realist imagery seems appropriate, of 'The State' doing good. The photographs never have a sense of agency in the images or the people themselves. They're always depicted as clients, or objects at the bottom of a pile. Very inert images. (Hall 1983/1984: 18)

This criticism of images that seem to present their subjects as mere representatives of some social or political problem has a long history. It is one of the many ways in which the efficacy of documentary has been questioned since the 1930s. At that time the idea that a photograph was a simple transcription of reality was widespread, but the desire to record the appearance of things accurately – the drive to realism – was under pressure.

It was possible at the time to find people and places about which very little was known, even in the societies of which they were part. The faces of the poor, the destitute, the criminal, and the workers had not been endlessly revealed. Hauling social destitution and human misery into the light of day seemed to be a radical act. The word 'documentary' was coined at this time and was quickly adopted because it seemed to encapsulate a version of realism that was more than the positivism of the past, but allowed film and photography to be both grounded in a world of facts and yet free to develop a practice that critiqued those facts and an iconography that pointed to ways of transcending the existing nature of things. Michael Denning commented acutely on this in terms of the United States:

> The documentary synthesis reoriented cultural history and criticism by changing the objects of study, making photography, and particularly the photographs of the Farm Security Administration, the central depression genre, and by turning critical debate to the formal and political issues raised by documentary: the problematics of capturing the 'real', the desire for the objectivity and immediacy of 'experience', the dangers of manipulation and propaganda. (Denning 1996: 119)

'The problematics of capturing the real' was, indeed, a central question. Towards the end of the twentieth century, Eric Margolis rehearsed one of the arguments of the 1920s and 1930s. That is, how much and what kind of 'reality' could photography reveal? Writing about images of miners he argued that photographs have emphasized the technical science of mining, but the social relations of production remain hidden:

> I conclude that photography constitutes an operationalized language incapable of expressing alienation or negation, potential, irrationality, alternative meanings, and so on. This has profound implications in a world where photographic images mediate so much of our experience. (Margolis 1998: 5)

This immediately invokes Walter Benjamin's celebrated critique of the New Objectivity movement in general and Renger-Patzsch's *The World Is Beautiful* in particular. He contended that the movement succeeded in making misery itself an object of pleasure, by treating it with technical perfection. He went on to quote Brecht on the inability of a photograph to express power relations or make visible the patterns of influence and control within social situations (Benjamin 1931). By the 1980s these older debates had been re-discovered, and questions of what we might mean by 'realism' or 'the real' were asked once more. This time, though, photography theory, criticism and practice was transformed by ideas coming from structuralism, psychoanalysis and literary theory. At the heart of this was the idea that photography did not address some external world of facts that had to be represented, but was central to the creation of that world. Photography was not regarded as a bundle of neutral technologies, but as a most important and shaping cultural practice, a practice that was deeply involved with other modes of representation. Photographs were seen not as innocent transcriptions of reality, but as densely coded and complex cultural objects that helped to create the real world that they pictured. Commenting on *Let Us Now Praise Famous Men*, Erika Balsom and Hila Peleg observed that the questions raised by that work are still our own and added that:

Far from any notion of 'fly-on-the-wall' immediacy or quasi- scientific aspirations to objectivity, such practices understand documentary not as the neutral picturing of reality, but as a way of coming to terms with reality by means of working with and through images and narratives. (Balsom and Peleg 2016: 13)

These debates about the relationship of photography to 'the real' are not simply played out in works of criticism; they are significant issues for practising photographers. For example, Stacy Kranitz spent many months over a number of years photographing mining communities in Appalachia. She is aware of the visual clichés of the past and is suspicious of the routines of photojournalism. She is uncomfortably conscious that her presence in a strange and damaged place means that she may be gazing at it with a colonialist's gaze. She says:

Representing place is a complicated series of negotiations. How can the photographer demystify stereotypes, represent culture, sum up experience, interpret memory and history? I create a fantasy world for myself. My perceptions and fantasies rival my desire to provide a realistic portrayal of where I am, especially because the idea of a 'realistic portrayal' is a fantasy. My work is about the tension between these two desires. If in fact they are two. Maybe they are just one. (Kranitz 2014)

Realism shot through with fantasy and desire is an intricate mix, but it is one that has often been present in depictions of mines and miners. Often 'realism' is seen as a synonym for 'positivism'. Kranitz reminds us that realism is more than an objective description of the world. It can be a complex concept that is formed by the subjectivity of the photographer as much as by the nature of what is in front of the lens.

Today, what might have been seen as 'straight documentary' is not the preferred mode through which workers are photographed. Perhaps the most significant figure who in recent decades has devoted a great deal of time to picturing workers is that of Sebastião Salgado. He has travelled the world taking subtle, complexly framed, but recognizably 'documentary' pictures of the labourers, the landless and the dispossessed. His work, however, is presented for the most part in monographs and on gallery walls rather than being in dialogue with writers in the pages of magazines. Salgado has turned his attention to global questions about the fate of the planet at a time of climate change and worldwide pollution.

Because he has photographed systematically and thoroughly, Salgado may be seen to have the archival ambition of producing a comprehensive study of a social group. This is, indeed, one of the ways in which some photographers have worked when they aspired to capture representative figures of an entire nation or culture. One thinks at once of August Sander who aimed to picture German people in portraits that sharply delineated them as individuals, but also showed them as representatives of their trade and class.

In 1955 Robert Frank produced what has become one of the most famous of photobooks, *The Americans*. This brought together a number of photographs that exemplified the Swiss photographer's take on post-war America. He avoided the kind of places that documentary photographers loved, but produced cool and ironic images of everyday life. Travelling widely across the country he recorded sites such as tract houses, empty freeways, drive in movies and bus depots. That he gave his collection of curious and off-beat scenes the magisterial title *The Americans* turned his laconic images into a critique of a society that was relishing its affluence, a world full of consumer goods, and its status as global power. Frank showed the other America full of anomic individuals inhabiting a nondescript or sleazy environment (Frank 1959).

A contemporary photographer working in this tradition is the Chinese photographer Liu Zheng. He grew up in the mining town of Datong and, after graduating from college, he was employed at Beijing's Municipal Mining Bureau and photographed miners in a number pits around the city. In the 1990s China was opening up to the outside world and transforming the nature of its economy and the relationship between the government and the people. At this time Liu Zheng was working as a photojournalist for the *Workers' Daily* newspaper, and he initiated a project to photograph Chinese people of all regions, classes and types. To this end he travelled extensively around the country and, it is said, produced more than 10,000 images. Out of this he produced his important book, *The Chinese*. Although photographing anyone that interested him from Buddhist Priests to drug traffickers, Liu Zheng is also very absorbed in traditional Chinese culture, especially in its erotic forms. Indeed, an apparent fascination with the bizarre and the deformed has frequently meant that he has been compared with Diane Arbus, whose portraits of American life also included the marginal and the grotesque. Among the more than a hundred pictures in 'The Chinese' one is of a mass grave, another of the naked body of a young man killed in a traffic accident and a third of a 'Mentally Handicapped Muslim Girl with her Nephew'. Alongside the dead are the dying, including those with freakish medical conditions. Clearly the body, marked by illness, age, deformation or touched with beauty is at the heart of the book.

One of Sander's photographs of a miner shows him blind and sitting on a bench next to a blind soldier. This is rather unlike his usual straight to camera shots of people revealing their trade or class. *The Chinese* has five pictures of miners. Only one has the signifiers of the trade, a black face and a hard hat with a cap lamp. He is, however, shot in a tight close-up with his face placed diagonally in the picture plane which lends it vitality. He is also smoking a cigarette, so that he has a rather louche appearance, rather than being a representative figure of hard labour. Then there is a picture of miners in a public bath house, one in the foreground looking up rather apprehensively at the camera while behind him a man vigorously soaps himself. The next shows two men in a communal bath. Their faces are still black from coal dust although they have washed their bodies. The photographer stands over them so that they are looking up into the camera with still, unsmiling faces. The second shows two naked men – one seated and newly bathed, the other standing and covered in the grime of the mine. The third is captioned 'Coal Miner, Arm Lost During Work. Datong, 2002'. Here a naked man stands among trees with his right arm circling his body. His left arm is missing from the shoulder, and the clear line of his body matches the line of the trees against which he is posed.

These are pictures of men with the lean bodies of those who do hard, manual labour and marked with the dirt of the mine. They are complex images of people who are sexualized and have subjective lives rather than being merely representatives of a particular trade (Zheng 2004).

Despite a high accident rate many Chinese people have moved off the land to work in coal mines. Over time, they have produced characteristic communities, and I want to turn now to look at the nature of coal settlements as well as exploring some kinds of coal gathering and collecting that take place in marginal places or in sites outside the world of organized mining altogether.

FIGURE 21 Liu Zheng, *Coal Miner, Arm Lost During Work, Datong, Shanxi Province, 2002.*

3 MINING COMMUNITIES: COAL CAMPS AND MINING VILLAGES

FIGURE 22 Walker Evans, *Interior of a Miner's Shack, Scott's Run Outside of Morgantown, West Virginia, 1935.*

In 1935 Walker Evans made a photograph entitled *Interior of a Miner's Shack, Scott's Run Outside of Morgantown, West Virginia*. In the foreground a young boy sits on the edge of a wooden chair while behind him in the gloom of the shack we can make out on the dark furniture a bizarre array of photographs, a cut-out Santa Claus together with magazines and newspapers decorating the walls. This is a complex, intriguing and emotionally charged photograph, but most miners' houses at the time had little to recommend them and photographs of them show rows of crude wooden buildings strung out in long lines. In 1925 a Report of the United States Coal Commission noted that:

> In the worst of the company controlled communities the state of disrepair at times runs beyond the power of verbal description or even photographic illustration, since neither words nor pictures can portray the atmosphere of abandoned dejection or reproduce the smells. (Quoted in Morris 1934: 88)

The challenge to describe the appalling conditions in which miners lived was taken up by novelists and became a major theme in works about the coalfields. Upton Sinclair's *King Coal* (1917), a novel of mining life, was based on the 1913–1914 Colorado coal strikes. It concerns a rich kid who, under an assumed name, finds work in a mine out of sociological interest. As in *Germinal*, which we discussed in the Introduction to this book, the main protagonist is an outsider who knows nothing of working class life, but opens our eyes to the techniques and politics of mining. He quickly learns that almost every aspect of life in the coal camp is under intense scrutiny from the managers and overseers. His fellow workers come from all over the world, and Sinclair tells us a great deal about the nature of the coal camp they inhabit. He describes it as a desolate and sordid place. Generally squalid and lacking all comforts, there were even worse quarters on this appalling site:

> There was a part of the camp called 'shanty town' where, amid miniature mountains of slag, some of the lowest of the newly-arrived foreigners had been permitted to build themselves shacks out of old boards, tin, and sheets of tar paper. These homes were beneath the dignity of chicken-houses, yet in some of them a dozen people were crowded, men and women sleeping on old rags and blankets on a cinder floor. (Sinclair 1917: 25)

Here Sinclair is describing the worst kind of American coal camp, but even those with better amenities were still entirely in the control of the coal operators. They supervised every aspect of life, from the company store, education, medicine and even freedom of movement. Private guards were employed by the coal owners but given state powers to arrest, dismiss or expel miners, using whatever force was necessary. The owners and operators of mines could also call on the militia to break a strike, and they drew up the list of suitable jurors to try court cases. Paid in 'scrip', that is, tokens that could be redeemed only in the company store, the miners existed in a state of servitude. This was not unique to America for although in Britain payment in company tokens was made illegal in 1863, it is generally considered to have gone on for some time after that. But even outside a hastily thrown up coal camp housing in mining areas has long been regarded as shoddy. Writing as late as 1948 Ferdynand Zweig described some villages in the north of England:

FIGURE 23 Russell Lee, Coal Camp, 1946. One of many coal camp images made for the Medical Survey of the Bituminous Coal Industry.

In many if not all places the surroundings are marked by ugly, brutal and bleak industrialism, as can be seen in some villages in South Yorkshire or in North Staffordshire, Durham, Northumberland or Lancashire. There are frightful landscapes dominated by the hideous grey slag-heaps which look like giant dustbins with dirt and filth all round. With their agglomeration of rows of stumpy houses, wires and rubbish widespread, ashes, mud and weeds, they make an awesome impression of soulless places deserted by God. (Zweig 1948: 43)

Most mining villages around the world began as hastily built accommodation designed simply as shelter for the early miners. Describing the situation in New Zealand, Matthew Wright noted that:

Mining communities often clung tenaciously to the landscape. Most mining townships emerged as clusters of temporary huts or tents, often paid for by the founding company and in most cases gradually evolved into more settled and multidimensional communities. (Wright 2014: 153)

Over time these places usually added the structures of more complex living. Schools, churches, shops and pubs followed and people began to acquire houses of their own. In the United States some people became very proud of their coal camps. Kopperston in West Virginia was founded in 1938 and became known as the nation's model coal camp with a variety of housing types and some amenities. While in the former mining towns of Artois and Béthunois in Northern France, you can see a range of well-built miner's houses from estates, terraces and garden estates. Bob Hayes writes of a boy growing up in a coal camp in the 1940s to 1950s. There were just 100 houses in the camp, together with a school, a pool hall, a church and a theatre. Despite the often-repeated claim that mining communities were egalitarian places, Hayes notes:

> This was a segregated coal camp as were all during that era, and even the bathhouse had separate shower stalls for the black coal miners. The funny thing was that when all the workers exited the mines each day, you could not tell a black coal miner from a white coal miner. (Hayes 2011: 289)

Zweig's description of mining villages as 'soulless' and 'deserted by God' may be socially accurate, but the influence of religion in mining areas cannot be overstated. Hayes came from a Methodist family, but he points out that not:

> All families living in the coal camps were not Methodist, of course. Other denominations included First Baptist, Free Will Baptist, Primitive Baptist, Catholic, Presbyterian, Pentecostal, Episcopalian, Church of Christ, Assembly of God, Holiness, and were represented by churches dotted all over the hills and valleys surrounding Appalachia and Big Stone Gap. There was also a synagogue located in Norton. (Hayes 2011: 388)

As well as containing churches, chapels and synagogues, what does a mature, traditional mining settlement look like? Often near a slag heap or two, they have wooden or stone houses strung out across a hill or huddled together in a valley. You'll find a few pubs, a library, a cinema and some sort of social centre. The rivers are dark with coal dust, and railway lines snake away from the pithead. On the pit surface is a series of buildings thickly covered in dust, with trams or wagons full of coal and timber that clank away throughout the day and night. The older the settlements, the less likely they are to have any trees left, for they will have been deforested decades ago. Now scrubland stretches out on all sides. But nearby there may still be arable land, and, for many years, mining was often a part-time or occasional occupation. Sowing and harvesting took priority, but when that was done, a man might go and dig some coal. All over the world this pattern of activity can still be found. Informal mines spring up, and in some countries this mixed economy continues, often to the despair of governments who would like to regularize the industry and bring it firmly within the capitalist modes of production, regulation and control.

But coal communities have always been difficult to categorize, once one has moved away from easy generalities. They are often set among beautiful natural countryside, but they are not rural places. They were among the earliest of industrial sites, but they frequently lack most of the infrastructure of industrial or urban regions. One useful category that has been proposed by the social historian Daryl Leeworthy, it is that of the 'industrial frontier':

> Simply defined, the Industrial Frontier is a region – most prominently a coalfield – which is removed from the influences of a metropolitan centre and which therefore constructs its own identity, its own cultural forms and its own institutions. (Leeworthy 2009: 57)

Not all coalfields fit into this category, but it helps to define some mining areas. What marks them all is the fact that they are dominated by the economic and cultural influence of a single industry. The famous solidarity of mining towns derives from the fact that the habits and mores of a male-dominated industry were imprinted upon the general culture. The isolation of coal communities undoubtedly led to their highly developed sense of community and subtle sense of being different from the rest of society. Of course, not all coalfields were to be found in curious enclaves. In nineteenth-century northern England, for example, coal was mined, but the cotton mills were also in full swing and there were large towns or cities nearby that offered a range of occupations and possibilities. But in places like South Wales that went from a rural economy with small populations to an extraordinary place with a large population, there was no competition and no models of other ways of life. Essentially, a colliery was sunk and a settlement grew up around it as people moved in to take up the jobs. And move in they did. From 1851 to 1911, 366,000 settled in the South Wales Coalfield. Most of the immigration to Wales was internal from other parts of the UK, but there were also many from Ireland and smaller numbers from Spain, France, Russia, Poland and Italy. A cacophony of languages and a mélange of cultural norms and habits eventually settled down into distinct communities.

The cultural life of mining settlements

The cultural output of some miners has often been celebrated. In 2017 three galleries of paintings by miners opened in the Durham town of Bishop Auckland. The Mining Art Museum houses works by painters such as Norman Cornish and Tom McGuinness. Its website describes the project:

> The Mining Art Gallery shares the experience of life underground and life in the community through eyes of the people who lived it. The Gemini Collection, now in its permanent home, is an artistic record of an industry and a memorial to a former way of life. The work displayed showcases the skill and creativity of these labourers and celebrates the achievements of the mining artists as a vital aspect of coalfield heritage. (www.aucklandcastle.org 2017)

Despite their poverty and long hours of toil, miners managed to take part in all kinds of leisure activities from pigeon racing to playing the banjo, from choirs to bluegrass, brass bands to writing poetry. In the 1930s a group of miners in Ashington in the north of England took some evening class lessons in painting and over the years went on to become noted artists. Known as 'the pitmen painters' they became celebrated and were the subject of a play by Lee Hall that received international attention.

While not every miner became an artist or an art lover, mining communities did manage to transcend the poverty and meanness of their surroundings and create social and cultural institutions of great merit. They built and ran libraries, cinemas, hospitals, convalescent homes, leisure centres and civic societies.

Jonathan Rose claims that 'The miners' libraries of South Wales were one of the greatest networks of cultural institutions created by working people anywhere in the world' (Rose 2002: 237). The libraries grew out of literary societies or temperance halls of the nineteenth century, although these were led by middle class people. Many miners had contributed to a fund to educate their children, but when school fees were abolished in 1891, they redirected the money to the miners' institutes. In 1920 the government set up the Miners' Welfare Fund, which put a tax on coal and gave the money to fund social amenities such as colliery baths, and libraries in the mining valleys. By 1934 there were more than a hundred libraries with an average stock of 3,000 books. They were housed in Working Men's Institutes, the largest of which offered evening classes, lectures, sports facilities and photographic darkrooms. They booked entertainers, and held dances and social events. Other coal regions had similar facilities, but these lacked the fervour of the Welsh libraries that were infused with the spirit of nonconformist religion, and the influence of Methodism which encouraged sobriety and emphasized the importance of learning and self-improvement. This extraordinary cultural life has often gone unnoticed, but by the 1930s the libraries had already begun to decline, partly because of an exodus of the young out of the valleys. Also, municipal libraries had grown in number and quality, and by the 1960s, most miners' libraries had closed. These cooperative ventures were designed to turn mining villages into ordinary places with some of the amenities of urban life. But they were fated to remain different, not least because the danger of the work meant that accidents and sometimes disasters were features of everyday life.

Disasters

If there is one thing that marked out mining communities, it was the ever-present fear of accident, death or disaster. In Britain at the start of the twentieth century over a thousand miners lost their lives in accidents every year. These were the deaths of individuals or very small groups, and they did not receive the attention given to the major disasters. Graham Greene, reviewing Pen Tennyson's film *The Proud Valley* in 1940, pointed out that:

FIGURE 24 The Maypole Colliery Disaster of 1908 had a death toll of 76 people and took place at Abram near Wigan. It was commemorated in this card.

No picture of a mining district ever seems to be complete without a disaster (we have two in this picture): the warning siren is becoming as familiar as the pithead gear shot against the sky – and that has joined the Eiffel Tower and the Houses of Parliament among the great platitudes of the screen; and yet a far worse tragedy in a district like this must be just inaction. (Greene 1972: 275)

In fact, many more miners died prematurely from industrial diseases such as pneumoconiosis, than from explosions or roof falls. But these were, and are, common enough. Today the death toll in coal mines around the world is still very high, although it varies from place to place. China accounts for most deaths in mining accidents, with the chances of being killed in a Chinese mine running at some thirty-seven times higher than in the United States. The country also holds the record for the worst mining disaster in history, which took place at the Honkeiko Colliery near Benxi in 1942. In an underground explosion that was caused by a mixture of coal dust and gases, 1,549 miners were killed. Ten miners were killed in 2017 in the United States, while in China, despite the government's actions to reduce the number of working mines, more than 7,000 people died every year in the four years leading up to 2000. The figures for India are almost as woeful. Some of these deaths receive a great deal of attention. In India, for example, in 2017, in the state of Jharkhand, a coal mine collapsed killing eighteen people and leaving more trapped under rubble in a strip mine. This event was reported around the world, and mining disasters have always had the potential to attract public attention. In 2014 *The Economist* magazine reported on a mining disaster in Turkey. An explosion led to an underground fire in a Soma mine:

As Turkey grieves over 301 miners who died in its biggest industrial disaster, in Soma on May 13th, another truth is sinking in. Most of them perished because of appallingly unsafe conditions, lax government inspectors and an operator that put profit first. Yet Turkey's prime minister, Recep Tayyip Erdogan, claimed that such tragedies were 'in the nature' of mining – never mind that methane leaks had been reported and ignored. (The Economist 2014: 45)

The response that mining is intrinsically dangerous and that disasters may be expected was quite a common one in the past. Britain's worst mining disaster was at Senghenydd in South Wales in 1913 when 439 men and boys were killed. A short photo essay in John O'Sullivan's *Photographic History of Mining in South Wales* takes us from images of rescue workers with a caged canary, via volunteers walking into the pit, to vast crowds at the pit head waiting for news. Then a coffin appears, just one among the hundreds that will be needed. Volunteers from the Salvation Army comfort the women waiting to hear the fate of their loved ones, until, finally, a vast funeral procession moves through the little town (O'Sullivan 2001).

Of course, newspapers carried stories of mining accidents, but one interesting way of spreading the news and commemorating the event at the same time was by the sending of postcards. These were often illustrated with portraits of the dead miners, sometimes in vignettes against a background of a pithead. We do not think of postcards as a medium for communicating serious and mournful news, but John Hannavy explains their use in disasters by pointing out that:

The idea of producing or sending a postcard to commemorate a disaster may seem odd today, but to those who lived in the early years of the last century, with no telephones, illustrated newspapers, radio or television, how else could pictures of important stories be disseminated? (Hannavy 2013: 128)

In his book *Coal Mining Disasters in the Modern Era,* Brian Elliott reproduces several of these photomontages. One composite image shows the victims of the 1908 underground fire at Hampstead colliery, near Birmingham, together with the men of the rescue brigade. He also prints a curious studio photograph of the only survivors of a disaster at the Maypole colliery. Dressed in their working clothes and holding lamps, the men gaze out at us while posed against a studio backcloth and draped curtain. One of the features of a disaster that is always present in films is the crowd of people that gather at the pithead to watch the activity as the rescue brigades go to work. Elliott has several of these, reproduced from the *Colliery Guardian.* Postcard publishers could work at great speed. Elliott observes that the Warner Gothard photographic firm was able to produce composite photographs within a couple of days. Here, too, are pictures of men helping the injured into ambulances, wearing the heavy breathing apparatus of the rescue brigade and carrying the dead on stretchers. One source was the *London Illustrated News* that used wood engravings to show fires at mines and, most dramatically, scenes when men were trapped below ground.

Disasters on film

It would be hard to find a novel about mining life that did not have an underground explosion with the surviving miners unable to reach the surface and depending on the work of rescue teams to save them. One of the most important of early mining films has exactly this scenario at its heart. Directed by G. W. Pabst in 1931, *Kameradschaft* was a realistic study of a mine disaster with wider political significance.

An underground fire and explosion in a French coal mine leaves men trapped below ground. The pit is situated on the border between France and Germany, and the story is based on a very serious industrial accident – the coal dust explosion that triggered the Courrières mine disaster of 1906. That was Europe's worst mining accident, with more than a thousand men killed. There was a concerted attempt to rescue anyone who might be alive, but the French lacked well-trained rescue workers and the outcome was extremely grim. In the fictional account of the event, the tensions between the French and the German people living close

FIGURE 25 A contemporary poster for Kameradschaft, 1931.

to each other is established early on. German workers looking for employment in France are turned away. A group of German miners cross the border to go to a bar where there is music and a good deal of drinking. The scene is played for comedy as the Germans speak no French and quickly find themselves in trouble. They want to dance with a young woman, Francoise, and when she turns them down, because she is tired, they think she has repulsed them because they are German. Another act of unfriendliness appears to have been made. The theme of borders is central to the film, as the post–First World War redrawing of boundaries means that the pit is divided between the two nations and simply separated by a gate.

When a fire breaks out on the French side of the mine, the Germans feel the heat through the wall while the French miners try to reinforce it to keep the flames at bay. When blasting is resumed at the pit, the fire bursts through the wall. There are very dramatic scenes underground as the fire rushes through the mine , while in exterior shots we see the mine on fire and people racing through the street. Francoise is leaving on a train for Paris saying she will never marry a miner, although both her brother and her boyfriend are miners and are about to be caught up in the disaster. The flames are visible from the German side, and little seems to be being done to rescue the trapped miners. In a frenzy of grief some women try to get into the mine to help rescue the men, and an old man sneaks in to search for his grandson.

There is more drama as the roof collapses, and in the German pithead baths the miners begin to say that they must go to help their French counterparts. Seeing what is going on, Francoise gets off the train and returns to the village. The German miners, some of them with breathing equipment, finally get to the mine and begin to rescue the survivors. The old man finds his grandson alive. On the surface night is falling and the moral of the piece is expressed: 'we are all united as miners'. The film contains most of the components of later works about mining: the drama of the initial accident; the human stories that connect us to the place; the tension involved in a rescue of this kind and the idea that being a miner transcends all other sets of identity.

Kameradschaft concentrated very much on the accident and the responses to it, but it was not very interested in a close study of the communities themselves. This, however, is very much the narrative line that runs through Carol Reed's film of A. J. Cronin's novel *The Stars Look Down* (Cronin 1935) that tells the story of a mining community through the conventional device of a single family, supported by one or two neighbours and friends. The film is framed for us by a narrative voice that, speaking over shots of miners emerging from the pit cage at the end of the day, tells us that this is the story of ordinary people who are:

Neither plaster saints nor romantic rebels. Yet these men and women are the backbone of nations, the stuff of human destiny; simple working people such as there are the world over in all countries and at all times.

This is an alternative version of the 'anywhere that is nowhere' of movies, and asks us not to see the setting as local and particular but as exemplary of the general condition of working people.

The opening scene sets up the major motif of the film. The miners are refusing to work a new seam of coal because they believe, on the evidence of the father of the hero of the film, that extant plans show it to be unsafe. The seam is a relatively thin barrier, behind which lie millions of gallons of water in old workings of the pit. The coal owner denies the existence of such plans and assures the miners that he has the agreement of the union to their working in what is a perfectly safe situation. They decline to work in this place and go on strike, but are eventually forced to return to work. Davy, the hero of the film, is studying hard for a place at the local university from where he intends to go into politics and hopes to become a Labour MP. His friend Joe scorns both mining and education but intends to better himself in a world of rather dubious enterprise. While Davy's ambition is for the betterment of his class, Joe is on the make. Both are initially successful, but Davy is lured into marriage with an unsuitable woman, abandons his studies and has to return to his village as a badly paid teacher. Meanwhile the unscrupulous Joe becomes increasingly successful and is responsible for persuading the owner of the mine to keep open the dangerous seam. The union, which refuses to intervene, is presented as being composed of people of irresolute will and with an eye to expediency.

The mine is flooded, and, in one of the best scenes in any of the films about mining, a small group of men, including Davy's brother and father, struggle to escape, but are trapped below ground. The rescue team works to reach them, and we have scenes of their fight for existence before another wave of the flood buries them all. Their deaths are presented as the result of feckless greed and a general indifference to the hazardous conditions that existed in coal mines. Of course, it is revealed that the plans are available and so the word of the working miner is allowed to triumph over the statements of owners and managers. Davy is seen as a man devoted to his class who through a lack of worldliness falls into despair, but who at the end of the film will continue his struggle.

It might seem that the theme of mining disasters as a plot device in films and novels has run its course, but as late as 2009 it still resonated in Indian literature. Mine number 3 is almost worked out, but a few workers are still employed there to strip out every last ton of coal. In a few months it will be closed. But a disaster strikes, and again, it is a predictable flood. Most of the men escape, but a small group are trapped, desperately seeking a way out as the waters rise. This is the situation in Sanjay Bahadur's novel *The Sound of Water* (Bahadur 2009). He is a former director of the Indian Ministry of Coal and would have known of the Bagdihi colliery disaster of 2001, on which the novel is loosely based. The book looks at the trapped men, led by an old and eccentric miner, but also at his family who stand to gain financially if he dies, and the managers who are charged with coordinating a rescue effort, but are more interested in their own positions and futures. The flood is not treated with the forensic detail of many mining novels. In fact, it is an extended metaphor for the inundation of evil and the various human practical, material and spiritual responses to it.

Moving and visiting

Migration has always been a feature of working class life and in the inter-war years 1919–1939, when the industry was in a depressed state, more than half a million people left the mining valleys of South Wales. This was far from the first large migration out of the valleys:

> In the 1840s, some two thousand Welsh, Scottish, English and Irish coal miners voyaged across the sea to the United States. These seasoned industrial workers found the New World a rural place with few smokestacks. Farmers dug coal in the winter where it outcropped on their land. They hauled the stuff to the buyer in carts or wagons, discontinuing the operation with the spring press of plowing and planting. (Long 1989: 6)

The influence of these immigrants was very great, and Priscilla Long goes on to claim that, in the mid-nineteenth century, the coalfields of the United States were essentially British, and that British capital also flooded into the coalfields of the United States. These flows of people meant that the vital characteristics of mining communities were shared across national boundaries. Wherever miners went they faced similar problems and created places with the same mix of cultural features. But they were always in thrall to the needs of the mine itself. Indeed, some commentators have observed that mining communities were structured so as to facilitate the smooth running of the industry:

> Mining families, centred around women, have functioned as vital elements in the organization of mining. In what might appear to be an incredibly uncanny fashion every detail of the immediate environment of the miners – their leisure, their homes, family relationships, wives and children – has served the structure of their existence, namely the provision of labour power with given skills in required quantities at the requisite times. (Allen 1981: 74)

On this account the transmission of skills and the constant presence of workers are facilitated by the structure of the community. Remote from other ways of life or countervailing pressures, mining could continue and its values be transmitted into the future. Sometimes, though, continuity and settled existence are regarded as one of the defining features of these places.

> Time, a sense of shared past, is of fundamental importance. The overlapping social ties of work, leisure, family and neighbourhood are re-enforced through

the sharing of a common past, and of family traditions of working in mining and often living in that community for two or more generations. Sons are destined to be miners and daughters the wives of miners. This is re-enforced by occupational homogeneity, and social and geographical isolation from the rest of society. (Bulmer 1978: 33)

It is, then, the interlinking of the vital collaboration of miners at the coalface with the patterns of social interaction in the wider society, together with a common understanding of the past, that makes for the singularity and special nature of coal communities.

A more dramatic way of defining coal communities is to see them (as some American commentators have) as 'sacrifice zones'. These are areas that have been permanently blighted by industrial activity, economic disinvestment or seriously degraded by exposure to pollution. People living close to strip mines or mountain top removal sites suffering the ill health and deleterious social effects generated by the poisoning of ground water and toxic waste are often put into this category. The sacrifice is borne by the miners, the landscape and those living in the region.

There have, of course, been many sociological accounts of life in mining camps and villages, but they have also frequently been visited by journalists and novelists. In the 1930s, stimulated by organizations such as Mass Observation, and the idea that we needed to know more about our own society, many literary figures travelled to coal country. Some investigators gazed upon the workers as though they were the inhabitants of a strange land; others sought to keep a 'proper' distance from the subjects of their research; many engaged in some form or another of masquerade, obscuring their real purpose behind an assumed identity. Some went with cameras, inspired with the documentary impulse to bring back irrefutable proof of the nature of things. Significantly, when they were in search of the working class, most investigators chose not to visit the booming car factories of Oxford or Coventry, but to seek out the 'real' working class in coal mining towns and villages.

In his 1930s tour of England the writer J. B. Priestley visited the coalfield of East Durham. Depressed by the ugliness and grimy air of the place, he observed that, despite the vital importance of coal to the economy, most people, including politicians and policy makers, knew nothing of the mining areas and asserts:

Most English people know as little about coal mining as they do about diamond-mining. Probably less, because they may have been sufficiently interested to learn a little about so romantic a trade as diamond-mining. Who wants to know about coal? Who wants to know anything about miners, except when an explosion kills or entombs a few of them and they become news? The railway and motor-coach companies do not run popular excursions to the mining districts. Pitmen are not familiar figures on the streets of our great cities. (Priestley 1934: 322)

One of the people who did want to know something about miners was George Orwell, who set out to visit not just a mining area but a working coal mine and produced one of the most influential accounts of life in a coal mining district. *The Road to Wigan Pier* was a Left Book Club publication, the first part of which was an account of life in a coal mine and a mining community (Orwell 1937). He lived in boarding houses, travelled to the pits and undertook the long walk in to the coal face, bent double and in great discomfort. He tries to give us a clear sense of the processes of mining, using his ability as a creative writer, but also employing considerable ethnographic skill. In part two of the book he uses his own life and experience to try to explain the relations between social classes. He locates himself carefully as a member of the 'lower upper-middle class' and analyses the prejudices, fears and snobberies of this and other middle class groups, as well as criticizing the attitudes, comportment and aspirations of the workers. In the first edition of the book, the publisher, Victor Gollancz, provided an introduction that was designed to 'correct' a number of Orwell's views on socialism, Russia and the working class. This, then, was often regarded as a problematic and curious book because it is partly a documentary account of life in a mine, backed up with statistics, and partly a visceral response to the smells, indignities and grubbiness of working class life. Then it becomes a treatise on the nature of class interactions, and the gulf that exists in most aspects of life between the author, the reader (assumed to be middle class) and the manual worker. As part of his account of life in Wigan, Orwell surveys the state of housing in this industrial area. He notes the damp and decay, the overcrowding, the lack of bathrooms and hot water, and, in mining areas, the constant threat of subsidence as the very ground on which they were built was undermined. What was called 'the housing shortage' was a well-discussed topic in 1930s Britain, and Orwell sees it as central to the state of the homes of the poor at the time:

> [And] that is the central fact about housing in the industrial areas: not that the houses are poky and ugly, and insanitary and comfortless, or that they are distributed in incredibly filthy slums round belching foundries and stinking canals and slag-heaps that deluge them with sulphurous smoke – although all this is perfectly true – but simply that there are not enough houses to go round. (Orwell 1937: 52)

Before coming to this judgement Orwell has looked around dozens of houses and even made a brief typology of them. Interestingly, he describes this as 'inspecting' the houses and, although he wryly acknowledges that anyone who turned up without any authority to inspect *his* house would be quickly ejected, he makes foray after foray into workers' houses. In fact, inspection of themselves and of their homes was a common feature of working class life. Sanitary inspectors, planners, medical officers and the police would all find it natural to scrutinize the property and behaviour of the poor. Despite the centrality of his book Orwell has been criticized for his attitude to the miners and their communities. For while he praised them as workers and emphasized just how hard and difficult their work was, he was contemptuous of working class life, and described the sordid nature of their existence in great detail. Patricia Rae has argued that the book was, essentially, a work of ethnography:

> One of the most striking affinities between *The Road to Wigan Pier* and the productions of contemporaneous ethnographers is its treatment of the British working class as a foreign culture – its characterization of the northern mining community as a second, colonized 'nation' or 'race' within Britain. Orwell repeatedly emphasizes the blackness of the miners' skins and their 'Christy-Minstrel faces' calling them 'negroes'. He describes his trip north in terms more commonly applied to journeys south: as the adventure of a 'civilized man venturing … among savages' (Rae 1999: 73)

In the 1930s many people attempted a literary ethnographic approach to the workers. They were regarded as paradigmatic proletarians, but, at the same time, in an era of Depression and mass unemployment, miners were also represented as beaten, hopeless and desolate. In terms of the dominant society they were also more or less invisible, and to find them required a journey and knowledge of the techniques that might tease them out of hiding. The magazine *Fact* had urged its readers to become anthropologists of their own lives by making the quotidian strange and looking on everyday life as if it were foreign. As an exemplary project of this kind they sent their own researcher, Philip Massey, to Nant-y-glo, a mining village in South Wales. Reading his account now one does not get the impression that he made a special sense of the place, but returned with a most conventional account based on interviews and the study of social conditions, wages, sickness rates and so on (Massey 1937). To penetrate such places did require travel, but it did not involve some of the more exotic strategies of the past. In 1902 Jack London decided to go into the East End of London in order to see what life there was like. He presented himself at Thomas Cook's agency, was refused a tour and so made his own arrangements. These involved changing his clothes for a disreputable outfit so that he might pass unobserved. Thirty years earlier James Greenwood 'the amateur casual' had spent a night in a doss-house disguised as a homeless man. Thirty-five years later George Orwell was to repeat the experiment going down and out and on the tramp in England. This form of investigation as a means of penetrating the secret world of work was to become significant later in the field of professionalized social science, but it was also used by some documentary photographers, who regarded it as important that they should work without the knowledge of their subjects, so that the shot could not, in any way, be regarded as 'contrived'. It also allowed them to step outside the complex set of political and social arguments that were in play. One characteristic response of photographers was to see themselves as part of the camera, merely recording what was in front of them. Bert Hardy, who worked his way up from being a delivery boy to a major photographer on *Picture Post,* always maintained that he had no interest in politics or in interpreting a scene, a task that he left to writers, while he simply photographed what was in front of the lens.

Humphrey Spender, commenting on the work he did in Bolton for Mass Observation, saw himself as an outsider who should never reveal himself to his subjects. Working for a number of magazines and for Mass Observation, Spender brought back complex images of working class life that were later thought to be exemplary images of the time. In one account of this work he described his procedure as allowing 'things to speak for themselves and not to impose any kind of theory' (Spender 1978: 7). What is interesting about the period is the plethora of conventional and novel means of investigation that were used. In addition to formal reports based on statistical investigation, there were varieties of journalistic reportage, films, photographs and newsreels. All involved some way of entering into the life of the people, of trying to understand the nature of the communities and of reporting back to an audience situated geographically and socially elsewhere. Not only were popular versions of the methodologies of social scientists appropriated by others, but the distinction between different genres of writing became ever more clouded.

Those writers, photographers, filmmakers and intellectuals who undertook the journey to meet the workers did so from a variety of motives and from different political standpoints, but most found it a difficult journey. Their intention was to bring back authentic accounts of life in these remote and curious communities. Some travelled with cameras, and many of those who did not went, as it were, with the metaphor of the camera – with what we might call a 'documentary gaze'. There is little doubt that many people in the mining villages resented being stared at as objects of scientific curiosity. An eccentric, but often insightful, writer described the travellers to the mining valleys of the Rhondda in characteristically forceful terms:

Many of the visitors to Rhondda settlements are … social investigators. They cannot help having that sort of spirit which inhabits the breast of the man who goes to see the animals in the zoo. Rhondda is distressed; her people are poor in a massed sort of way; Rhondda presents an incomparable site for an examination into social problems. Thus Rhondda has her sociologists who cannot help looking upon a Rhondda man or woman as part of a statistic or a useful source of information … the whole spirit of such a person is low and mean. (Edwards 1937: 173)

Even the most dedicated amateur anthropologist, determined to maintain the stance of impartiality, is likely to be daunted by such a response. Some reporters of the time, however, were either unconscious of, or chose to ignore, the difficulties of encountering their subjects. Their journeys were undertaken with the desire to 'see for myself' and the travellers who made them were convinced that the evidence of their senses would be enough to sustain them. Striving to attain the condition of the camera left such writing open to the weakness of the camera: that it could show suffering, degradation, despair, but could do nothing to illuminate the causes of these woes. Power and causality are difficult to express through the camera eye, as are collective struggle and resistance.

Artisanal mining

The communities that were under investigation were settled coal-mining villages whose inhabitants worked in deep coal mines, but there are many other kinds of mining whose workers do not live in conventional mining camps, towns or villages. Around the world coal is gathered in all kinds of marginal lands and from makeshift communities. One of the photographs in Ilan Godfrey's collection *Legacy of the Mine* shows two women and a child in South Africa. They are standing by wheelbarrows, one of which is full of slaggy coal. All three are gazing at the camera, and the sun casts harsh shadows on the ground. The picture is titled *Sylvia, Angel and Setty, disused coronation colliery, Likazi, Emalahleni (Witbank) Mpumalanga, 2011* and there is an explanatory caption:

Sylvia Mlimi, Angel Mona and Setty Mndawe with dirty coal collected on Coronation Colliery. Collecting coal is a daily necessity and is fraught with danger, especially when the ground collapses while digging. People from the Likazi informal settlement have died as result. (Godfrey 2013: 136)

These people are doing something that is as old as mining itself. For while coal mines have become more mechanized, larger and increasingly technically sophisticated, the world is full of people who dig for or gather coal in dangerously informal ways. These processes are collectively known by the rather charming title of 'artisanal mining'. This phrase covers a range of activities from farmers who simply turn over outcrops of coal on the surface of their fields, to people who sink shafts and burrow tunnels beneath the surface of the earth. In other words they scrabble for coal free from the body of rules, safety procedures and legal constraints that regulate the work of formal mines in most countries. In some places children routinely work alongside other members of their family, but hard statistics on this or, indeed, on the scale of accidents and injury are hard to find.

Because artisanal mines are small in scale it would be a mistake to assume that they are unimportant. For instance, one study in 2001 suggested that there were 6 million artisanal miners in China and that they produced about 11 per cent of the world's coal output, making the sector more productive than the total output of the formal mines of Australia or India. The Chinese government has long been anxious to close down these mines for social and environmental reasons, but they are often a vital source of income for subsistence farmers. Some are

FIGURE 26 *India, Jharkhand, Jharia, Children Collect Coal from an Artisan Mine.*
Photographed by Joerg Boethling. Original in colour.

regulated by the state so that, for example, children are not employed in them, but in remote areas it may well be that people pay little attention to the rules (Gunson and Jian 2012).

There are many ways of gathering coal. Perhaps the simplest is that of scavenging useful nuggets of burnable coal from slag heaps. This is what the women in Ilan Godfrey's photograph are doing, but this picture is in a long line of photographs of coal gatherers. In the 1930s the miner was not portrayed in Britain as a worker in heavy industry, but as emblematic of the wretchedness of unemployment, dereliction and human waste. Scrabbling for coal characterized this state, and there are many photographs of coal pickers. One reproduced in the first edition of Orwell's *The Road to Wigan Pier* shows a long line of men picking coal from a slag heap while below them the colliery is in full swing with the winder in action and steam trains hauling coal. There is a kind of collective effort and camaraderie in this image that gives the onerous task a friendly social aspect. Many pictures, though, strip away any kind of collectivity, but show a single, dark, bent figure wearily heading for home with a hard-won load.

The practice of coal searching was discouraged or forbidden by the mine owners or managers on the grounds that it was very unsafe. Miners, however, believed that this was simply a way of trying to force them back to work during strikes. Journalists and novelists did not miss the sad irony of formerly productive miners being reduced to scavengers of waste on the heaps they had created in the first place.

Bootleg mining

In the 1930s a particular kind of artisanal mining took place in the United States. It was known as bootleg mining and consisted of unemployed miners who, at first, harvested coal from existing mines in order to heat their homes. They then went on to sell coal in quite large quantities. Eventually some 10 per cent of anthracite coal reached the market from the bootleggers. If they were caught juries refused to convict them and the public was largely on their side. The coal owners fought back and tried to wipe out the bootleg miners, but they survived for decades. Indeed, Marc Brodzik's 2008 film *Hard Coal* is subtitled *Last of the Bootleg Miners*. Government policies citing health and safety concerns helped to put the informal mines out of business, but perhaps more important was a fall in the price of coal on the market that rendered the practice uneconomic. But the film emphasizes the culture of mining as at the root of the bootleg trade. The miners in the film describe themselves as 'the last of a dying breed' and claim that digging for coal is 'in your blood'. This idea that mining is more than a commercial trade, that it is an inherited passion that flows from generation to generation, runs through many discussions about the industry. This feeling is particularly strong in the unregulated, small-scale sector, which takes on almost a domestic note. One woman in Brodzik's film, talking about a backyard mine, expresses this very cogently when she says, 'He dug this himself. Him and the kids are the ones that done this'.

Across the world the informal collection of coal still goes on as was illustrated in a piece in the *China Times Press* in 2010 titled *Wumeng Miner*:

> During a photographic trip to the Wumeng mountains in Northeastern Yunnan Province, photographer Geng Yunsheng was flabbergasted to see a mine worker climbing out of a mine shaft less than a meter high. Clearly exhausted and sweating profusely, the man held a kerosene lamp in his mouth as he dragged a big bamboo basket with over two hundred kilogrammes of coal. (Parr and Wassink 2015: 322)

Geng Yunsheng made many trips back to the region and systematically photographed vernacular miners at work, resting, and playing cards. He also recorded the surrounding environment.

The Chinese film director Wang Bing also looked at the curious world of artisanal coal in his 2009 movie, *Coal Money*. It is set in Inner Mongolia and follows a group of truck drivers as they load up with coal at a private opencast mine in Shanxi Province and head for the port of Tianjin. The truckers are all petty entrepreneurs, struggling to squeeze as much money as possible out of the coal they are hauling. And it is not easy. They sleep in the cab, drive through dust and eat at sleazy roadside cafés. They pay for licenses and bribe their way through roadblocks. They have to negotiate and cheat to make any profit from their coal. They can haul about 38 tons of the stuff at any one time, but the price they can negotiate depends on the quality of the coal, and the buyers are always ready to declare it full of stones and more or less worthless. The drivers use mobile phones to try to find the price in other places to firm up what they can reasonably ask for. At just fifty minutes long it is a quite difficult to follow the action. Wang Bing's genius is for the slow unfolding of events and the long study of people. His other films run for eight to twelve hours! He has said of this film:

> Coal Money is an incomplete project. We shot a lot at the time. But it was done for a television programme in Europe which only gave me a 50 minute slot. The producer, a French company, actually understood the problem. They asked me to make a complete version afterwards, but I didn't have time to go back and work on it again. Within the 50 minutes it wasn't easy to narrate a coherent story. It is not a completed work. (Bing 2013: 117)

Wang Bing's coal operators are dedicated to travelling huge distances and taking a risk on the possibility of making a good profit. But small-scale mining is not always the sole source of income for the miner, and may be quite marginal. *The Bay Boy* is a coming-of-age movie set in Nova Scotia in the 1930s. Kiefer Sutherland plays a teenager in a Catholic household now poor after losing their business in the Depression. They get by as best they can, they take in boarders, his mother makes pies and cakes for the local shop, and his father descends into the cellar to dig out some coal. This micro-scale mining passes without comment in the film. Although they live in a mining town, the miners themselves are little in evidence. They live in a coal camp distant from the heart of the town, and invisible in the daily life of the place. But the state of the mine, the days when there is no work, strikes and accidents are constant topics of local conversation. In a voice-over at the end of the film Sutherland says that he left for university and 'I said goodbye to the mining town at the end of the earth and never returned. Only in memory'. This melancholy line resonates with the idea that what is being left behind is a place 'at the end of the earth' not simply geographically, but also in social and political terms. Mining, Sutherland is reminding us, is rarely to be found in the busy centres of commerce, surrounded by social and cultural organizations, but always at the margins in remote spots often bounded by deep valleys, mountains or the sea.

Free miners of the forest

The feeling that people have the right to dig for coal in their own region, regardless of property laws, has very occasionally been sanctioned by government, as in the curious case of the 'free miners' in England's Forest of Dean. Iron and coal were dug here in Roman times, and the population gradually gained certain rights. These were codified in the *Book of Dennis* in 1612 that spelt out the long-standing rights of the Dean Free Miners. Essentially this gave them a monopoly on the mining of coal and iron in the forest. Over the centuries the miners were commandeered to take part in military campaigns and became noted as archers and sappers. By the time of the Industrial Revolution, when demand for iron and coal was at its height, there was an attempt to rationalize the system. However, the Dean Forest Mine Act corroborated the existing arrangement and the Free Miners kept their privileges, which would be administered by a Deputy Gaveller, a post that still exists. You may become a free miner if you were born and are living within the hundred of St. Briavels. In addition, you must be aged over 21 and have worked for a year and a day in a mine within the hundred. This long-established, arcane arrangement may be seen simply as a historical curiosity, with its archaic language, curious terminology and feudal system of rights and obligations. But in this pre-capitalist system we may also see it as an organized and legally sanctioned version of the spirit of independent coal miners around the world. When a contributor to Brodzik's film *Hard Coal* says, 'we were raised in the coal business and we mine coal', he is invoking the spirit that motivated the free miners. For centuries people have felt that they have a right to dig in the earth beneath their feet and have resisted the demands of governments, coal companies and safety experts to control the land and turn them into wage earning labourers.

Sea coal

It is this spirit that also inspired the sea coal gatherers who scrabbled for nuggets of coal on ocean shores. This is one use of the phrase 'sea coal' but, confusingly, sea coal is also what the British called all mined coal until the seventeenth century, in order to distinguish it from charcoal. Barbara Freese says that the reason for the curious name (since coal is decidedly found deep in the earth) is unclear, although her preferred explanation is that 'since most mined coal had to be shipped by sea to distant markets, it became inextricably linked to water in the minds of those who burned it' (Freese 2005: 21). Another explanation she offers is that sea coal was so called because the North Sea carved out coal from outcrops and dumped it on the beaches. This, indeed, is how the sea coal gatherers worked by swiftly collecting the coal at low tide as the waves retreated leaving lumps of inferior but burnable coal exposed.

In Chloë Mercier and Edward King's film *Seacoal* we follow Joe Smith, a man in his 70s, as he leads his horse and cart across the beach and shovels the lumps of coal that are piled up at the water's edge. In this case the coal comes from a colliery tip a mile or so along the coast. The waste is swept out to sea and then returned to the beach with the tide. So, in some ways this sea coal collecting is just an elaborate form of slag heap searching, albeit one that was once very much more profitable. It was especially so just after the Second World War when the beaches were open to the people again. Coal was a vital and scarce commodity, and almost everyone heated their homes with it. By 2006 when the film was made only a few people were left combing Lynemouth Beach in Northumberland. In the film Joe recalled a time when they might garner 2,000 tons of coal in a weekend and more than eighty men and horses were to be found at the work. But times had changed; the pits had closed and no new material was tossed onto the slag heap. And if the supply side of the equation had changed, so had the demand for coal. People now lived in centrally heated houses and there was little call for smoky, low-grade coal that could be burnt on an open fire. Joe scratched a living by working the beach in all weathers and labouring hard to fill his cart and sacks. In the last days of his working life, however, he was able to gather coal without much competition, and it was the fierce rivalry between the many collectors that made the trade a tough one to practise in the past.

Chris Killip was born in the Isle of Man just after the Second World War and has worked as a photographer since he was 17. His subject has usually been working people who are trying to live full and rewarding lives in a world that is collapsing around them. In an essay in Killip's 1988 book *In Flagrante*, set largely in the North East of England, John Berger and Sylvia Grant noted:

Today the shipyards are silent, many of the mines are closed, the factories shattered, the furnaces cold. The tragedy of this has little to do with new technology as such, or with so-called post-industrialism. It stems, it bleeds, not

FIGURE 27 Sea Coal gatherers dig into the sea at Lynemouth in Chris Killip's 2011 collection *Seacoal*.

from the fact that science has discovered electronics, but from the fact that everything which constituted the loves of those living here is now being treated as irrelevant. (Grant and Berger 1988: 87)

When an entire industry is abandoned it takes with it not only the chance of a good economic livelihood, but also all the known and familiar components of life. The taken-for-granted, inherited patterns of existence are fragmented, so that everything from the way the family functions, the means of getting a living, the way communities live together, the sources of leisure and pleasure are all changed or destroyed. This rupture with the traditional way of things means that every aspect of everyday life may become strange and challenging, and the customary way of life for working class people was often associated with a particular industry. Working in an industry, people entered into a particular culture and took on a recognizable and respected identity. This is especially true of mining communities, partly because of the exceptional need for solidarity at the workplace, but also because they were and are often in geographically remote places. It may also explain why bootleg miners will happily dig a hole in the back yard and labour in a dark and dangerous place for small rewards. The activity structures the present by evoking the past and gives meaning to a way of life that appears from outside to be grim and obsolete. In the same way, the sea coal workers are practising traditional skills, but they are also working in a highly competitive trade and doing it for cash in hand and the chance to make a living. In 2011 Chris Killip published a book of photographs of the miners of Lynemouth Beach, simply titled *Seacoal* (Killip 2011).

In a short introduction Killip describes the difficulty he had in approaching the men who garnered the coal in Northumberland. In 1976 he was driven off the beach by angry coal gatherers. As he uses a large plate camera, mounted on a tripod, he could hardly be successfully surreptitious. Two years later he tried again with exactly the same result. So he sought out the men in the one pub in the place. He writes:

I announced myself as the person they tried to kill today … I ordered drinks and told them who I was: a photographer from the Isle of Man now living in Newcastle … I was interested in everything about the region and knew it quite well, although I had never seen anything like their seacoal beach, which is why I wanted to photograph it. I swore on my life I was not from the social security, the dole, the police or the tax office. I didn't belong to anybody. 'I am who I say I am. I will make certain that nobody comes to any harm because of my photographs. Can you just stop trying to kill me and let me photograph on the beach?' (Killip 2011: n.p.n.)

This appeal was unsuccessful, but a single contact, a man of significance on the beach, ironed out his difficulty and he was finally accepted. As the film about Joe

Smith makes clear, the sea coal miners lived in a settlement of mobile homes and in 1983 Killip pitched his own caravan on the site and lived there for more than a year. The result is a book full of superb photographs that add up to a moving portrait of a way of life that no longer exists. The black-and-white images are flooded with the sense of a landscape. We feel the coldness of the place in early morning light as an occluded sun struggles to be seen. It is a world of greyness and a world of work. The collectors, up to their knees in freezing seas, haul up spadefuls of small coal into carts drawn by their patient horses. We can feel the weight of the work and, through the gestures of the body, understand something of what it feels like to be shovelling coal in the icy morning air. The people and the landscape are moulded together. But these are also images of family life. Children help with loading the sea coal, but they also ride ponies, lark about, cuddle up to their parents, have a haircut or hold a dog, a rabbit or a frog. We see the caravans of the coal camp and the long curving stretch of the beach that leads to the distant mine that produces the coal. These photographs are usually categorized as 'social documentary', and certainly they have more than a little in common with classic proponents of the genre, but Killip was influenced early on by Walker Evans and Paul Strand, both photographers who resist easy categorization. The photographs in *Seacoal* have no captions, and they need to be read together rather than as single images. They are documents, in that they tell us a good deal about the work and the nature of the life they are drawn from, but they are not intended to be exemplary of any kind of social problem. Nor do the people have the wary look so often to be found in the subjects of documentary photography. They are relaxed and easy in the company of the camera. Sometimes the pictures resemble family photographs in their apparent spontaneity; at others they are more aesthetically charged. For example, a horse trots past a caravan in a grey light and is reflected in a large puddle. These, then, are complex images that can be read in several different ways. Now that this world has disappeared, the photographs take on new meanings, as they exemplify an extinct trade and a vanished way of life. Reviewing the book Clive Dilnot argued that

> the images that Killip offers are neither a record nor a document. They are both less and more; less because they are not as evidential as might at first be thought; more because instead of an ethnographic report, the book is a gathering of portraits. It would not be unreasonable, in fact, to call it a last collective portrait of the working class. (Dilnot 2012)

This is problematic, not least because the sea coal gatherers occupy a rather ambiguous class position, but it is certainly an important work and Dilnot is right to draw attention to the collection of portraits it presents. Around the world, especially in vast countries such as Africa and India, informal kinds of mining persist and are likely to carry on for some time. China is making a great attempt to outlaw artisan mining, and it seems that this has led to a fall in the number of fatalities. But if the fate of this kind of mining is in doubt, so is the future of mature coal communities as the drive against coal, and the pollution it generates, continues.

The future of coal communities

In 2016, while running for the presidency of the United States, Hillary Clinton published a 'Plan for Revitalizing Coal Communities'. This began by acknowledging the contribution mining had made to the American economy: 'The hard working Americans who mine, move and generate power from coal put their own health and safety at risk to keep our factories running and deliver the affordable and reliable electricity we take for granted.' This encomium was needed because Clinton had previously incurred the ire of miners by pledging to close down the mines and abandon coal as a source of fuel in the United States. The plan for the communities was designed to rescue them from the economic fallout of meeting 'the climate change challenge' by moving to sources of clean energy. The plan involved spending $30 billion on a range of activities designed to stimulate local economies. The strategy was intended to free up already existing assets, for, as the paper says, 'From Appalachia to the Uinta Basin, coal communities have rich human and cultural capital, diverse natural resources, and enormous economic potential'. The investment was targeted at developing infrastructural projects, improving health and education, repurposing the landscape, as well as increasing investment and developing an entrepreneurial ethos.

We will never know how successful such a scheme might have been, but it is only one among many strategies that have been proposed to comprehensively revitalize coal communities when the coal has run out or been abandoned. Throughout the United States and Western Europe there have been numerous plans to develop infrastructure, reclaim land for parks or retail complexes, build tourism or establish heritage centres. Around the world, there have been many projects that attempted to regenerate erstwhile coal communities. Although each of these takes on a national or regional complexion, they face some of the same problems. The landscape of mining country has been despoiled, often over many years. The population has low educational attainments, and many lack the skills that are called for in modern life. As young people often leave to try their luck elsewhere, the population is ageing. Ex-miners, now getting old, are afflicted by a range of industrial diseases, and are in need of good medical care. These difficulties would be recognized in the Pas-de-Calais or in Appalachia as much as in Scotland or England. Indeed, in his 2016 documentary film, *After Coal*, Tom Hansell looks at the connections and similarities between individuals in Kentucky and South Wales who were trying to find new ways of life and attempting to revitalize their communities.

Former mining towns and villages, however, are not simply locations in need of social work projects and medical intervention. They still function as communities, and many of the values that derive from their most vibrant days are still to be found. For instance, the people in these communities always declare that they are very supportive of one another. The claim is that the camaraderie, intrinsic to the patterns of work in the collieries, has long flowed out of the mine into the wider world. This is said to be particularly true of townships and villages that are remote from large centres of population. Of course, the sense of remoteness is social as well as physical; they were not places people came across on a journey or visited for pleasure. Also, the industry that once gave them their prosperity absorbed most of the male workers, so there never was a diverse pattern of commerce and industry on which they might draw.

4 FOG, SMOG AND POLLUTION

FIGURE 28 People are led through the foggy streets by men and a boy carrying lighted torches. *The Illustrated London News*, Vol. 10, 1847.

Fog in London

In 2017 a post-Brexit poster designed to lure city firms and workers from London to Paris read, *Tired of the Fogs? Try the Frogs*. This wordplay on two stereotypes was clearly a simple joke, but it is curious that London is still known as a foggy city. The last real fog there was in 1962 and that came after almost a decade of cleaner air. Half a century later it seems that the city and fog are still yoked together, at least in advertisers' minds. Yet this is not really surprising, for London earned its foggy designation over several centuries, and treated its pollution with remarkable insouciance, even at times with pride. Clean air movements existed from the nineteenth century, but effective legislation was not enacted until the 1950s.

In 1954 Gerard Bryant made a film for the Gas Council with the memorable title of *Guilty Chimneys*. Its purpose was to demonstrate the advantages of gas, described as a clean, modern fuel, over dirty, polluting and old-fashioned coal. Modernity was the key theme of the piece. It began by looking at air pollution, and it wanted the audience to understand that it is not merely dirty and unpleasant, but is 'an invisible killer'. Shots of smoke belching from chimneys introduced us to a city shrouded in what they call 'smoke fog'. In a documentary style we are shown X-rays of lungs that look grimly dark, and fragments of corroded buildings that have been eaten away by acid rain. The film also used white-coated experts to explain the science of pollution, and we learn that 'smoke fog' cuts out vital vitamin D, which we absorb through the skin in sunlight. This shortage leads to a weakening of the immune system.

The film told us that not only is coal deleterious to our health and to the preservation of city buildings, but it is also grimy and dirty and thoroughly out of keeping with the shiny white appliances we should aspire to own. To point this up they presented a man caked with dust heaving a sack of coal into a house where the brightly burning open fire was described as the main source of pollution, 'worse than industry'. A sonorous voiceover tells us that 200 million tons of coal is burnt every year. The film proselytized for gas, coke or other smokeless fuels to replace the burning of coal. Now we see a clean, modern kitchen fitted out with up-to-the-minute equipment and home to a housewife who moves among the knobs, jets and dials of her cooker with happy ease.

But the heuristic theme of the film continued as the viewers were taken to a public lecture on air pollution. Before an audience of ordinary folk the speaker spelt out the problem: two and a half million tons of smoke are discharged into the atmosphere every year. He argued that industry could cut back on emissions by replacing coal with coke, while domestic fires should also use smokeless fuels. He also called for cities to establish smokeless zones, which would ensure no, rather than fewer, emissions of smoke. This mundane film raises many of the issues that not only worried people at the time, but that had been debated for hundreds of years. To the medical case for an end to coal fires, it added the theme of modernity as exemplified by gleaming white-goods. In the mid-1950s there was a spurt of consumerism in Britain. The immediate post-war economy was devoted to the recuperation of industry, but a decade later, wartime rationing was finally abolished and a new age of personal consumption began. Now coal was not just deadly; it could also be portrayed as outmoded, belonging not to ease and domestic comfort, but to a bleak era of hardship and toil that was fortunately now over.

The film was also well timed because in 1952 London had suffered one of the worst fogs in its recent history, with five days of almost total darkness and noxious smells. It has been estimated that some 4,000 people died as a result of the smog and many thousands were made ill. It brought the city to a standstill and gave ammunition to the environmental groups that had been struggling for many years to convince the authorities to adopt a clean air policy. It was, however, not the first time that people had been urged to switch from coal to other fuels. The Smoke Abatement Exhibition, which was held in London in 1881, and visited by around a hundred thousand people, showcased new, clean appliances for domestic and industrial use. The exhibition moved to Manchester, where it attracted a similar numbers of visitors and urged the citizens to give up their addiction to coal (Thorsheim 2006).

Both London and Manchester were famous for their fogs. In the capital they were peculiarly yellow in colour from the high sulphur content that leaked from the bituminous coal. From this the frequent fogs became known as 'pea-soupers' and 'London Particulars'. Although there is a centuries-long tradition of fogs in the city, they became the product of smoke mingled with fog from the 1820s and reached their apogee in the 1880s when both the industrial and the domestic uses of coal were at their height.

The amount of coal used in the capital had long been a matter of wonder. As early as 1726 Daniel Defoe, on his tour of Britain, was astonished at the volume of coal waiting to be shipped:

> From hence the road to Newcastle gives a view of the inexhausted [*sic*] store of coals and coal-pits from whence not London only, but all the south part of England is continually supplied; and whereas when we are at London, and see the prodigious fleets of ships which come constantly in with coals for this increasing city, we are apt to wonder whence they come, and that they do not bring the whole country away; so, on the contrary, when in this country we see the prodigious heaps, I might say mountains, of coals, which are dug up at every pit, and how many of these pits there are; we are filled with equal wonder to consider where the people should live that can consume them. (Defoe 1726: 356)

But consume them they did, and by the 1870s, 11 million tons of coal were being imported to London every year. Over the next two decades the London fogs became increasingly frequent, more and more deadly, and an important component of popular culture, especially in writing about crime and deviancy. Cities had long been seen both as places of light and civilized life and as centres of criminality where dark deeds flourished in the gloomy, fog ridden streets. At the same time, blazing coal fires were never simply a way of getting warmth into a house, but were the very signifiers of happy family life. Hearth and home were indissolubly linked. So the obscuring fog that rendered vision ineffective and the guardians of law and morality ineffectual acquired a symbolic status. Christine Corton argues:

> Fog became a symbol for the threat to the clear outlines of a hierarchical social order as it dissolved moral boundaries and replaced reassuring certainties with

obscurity and doubt. The threat posed was not just a social threat, however, but also an individual one. It allowed the criminal, the deviant, and the transgressive to roam the streets unhindered and unobserved. And it placed them in a position to impose their authority on the respectable. (Corton 2015: 86)

So, in addition to undermining the lungs and choking the nose with appalling odours, the smoke fogs rendered ambiguous the social norms and moral certainties of the city. By the 1880s fog had become a potent subject in popular fiction. For centuries it had been associated with drink, the breakdown of moral order and the rise of dissolute characters. The famous serial killer known as Jack the Ripper was active in Whitechapel in 1888. The perpetrator was never caught and still inspires people to try to solve the mystery of his identity. A general fear of the poor and the working class was amplified into a terror of darkness, of gloomy streets and obscuring fogs. In 1887 Sherlock Holmes came on the scene and he was to have a most profound effect on British popular culture. While Holmes spent little of his time in a London Particular, he is frequently pictured as swirling through the fog in a Hansom Cab. And if he was not actually moving through a cloud of pollution, he was cutting away the crime and depravity that were its moral and social equivalent. The Ripper murders were particularly potent and inspired many works of fiction. In 1914 Mary Belloc Lowndes published her novel *The Lodger*, the tale of a landlady who thinks her lodger, a quiet, religious man who has saved her and her husband from penury, might be 'The Avenger', a serial killer who the police are chasing (Lowndes 1914). The book was clearly based on the widely known story of the Ripper, but in the changing character of the main protagonist, it also reminds us of Stevenson's *Dr. Jekyll and Mr. Hyde*. Fog pervades the London streets and once again conjures up an atmosphere of malignancy and danger. The book was a best seller and continues to be in print (Lowndes 1914). The story has been filmed several times, first by Alfred Hitchcock who made a silent movie based on it with the title *The Lodger, A Story of the London Fog*, and, most recently, by David Ondaatje in 2009.

Coal gas

In the early years of the nineteenth century an efficient system of street lighting was established and the fear of darkness began to be dissipated, but when fog rolled in it returned the streets and alleyways of the city to obscurity. Before the introduction of street lighting, cities were dangerous places at night not simply from the possibility of attack, but, more usually, from the inevitable accidents that happened as people blundered about the dark streets. The oil lamps in use were easily extinguished and gave a very uncertain light. From the very start of the nineteenth century people were demonstrating the virtues of gas lighting, which were established in Paris in 1801 and in London six years later. Cotton mills were illuminated from 1806, and in this artificial light the working day could be considerably extended. In the next decade the gas meter was invented, as was the gasholder. By the 1920s oil lamps had given way to gas throughout inner London. This was hailed as a great triumph, and gaslight was eagerly taken up by theatres and music halls as well as illuminating the streets. However, Roger Ekirch points out:

> Whether lives were consistently enriched by pushing back the darkness is less evident. Besides emitting a nauseating smell, coal gas was a growing source of pollution. Critics complained that the light was too harsh. Shift labor expanded in factories, as did industrial surveillance by owners. If night became more accessible, it also became less private. (Ekrich 2005: 333)

Many people, though, were full of enthusiasm for the new street lighting, for although some kind of public lights existed in cities, not until the coming of gas was the world transformed. In Paris thousands of lights illuminated the new boulevards, while the world's first department store in 1852 introduced the art of window dressing, so that commodities were enticingly visible to the crowds of people who promenaded along the streets late into the evening. The first age of the spectacle had arrived.

The dustmen of London

In his magisterial survey of London life and labour, Henry Mayhew interviewed a dustman and pointed out that Londoners used 3,500,000 tons of coal every year (around eleven tons per household), and he looked at the disposal of the residual ash and dirt. With his usual punctiliousness he describes the organizational structure involved, which centres on the work of the 'dust–contractors', people who have all the equipment necessary to remove the waste and a place on which to store it. There are, he estimates, some eighty or ninety such people and they 'are generally men of considerable wealth'. In his novel *Our Mutual Friend*, Charles Dickens created a character who

> grew rich as a Dust Contractor, and lived in a hollow in a hilly country entirely composed of Dust. On his own small estate the growling old vagabond threw up his own mountain range, like an old volcano, and its geological formation was dust. Coal-dust, vegetable- dust, bone-dust, crockery dust, rough dust and sifted dust- all manner of dust. (Dickens 1865: 30)

The waste, Mayhew tell us, is sifted and many barges carry the fine dust to places where it is used to improve the quality of wet and marshy ground. This once-prosperous trade had declined significantly by the time of Mayhew's account, but the dust was still mixed with clay to make bricks. At the metropolitan dust centres both men and women worked on the sieving process. People spent their lives in these centres 'in the midst of effluvia most offensive', yet (he assures us) remain healthy and strong. However, he describes both the men and the women who carry out the trade as largely illiterate, irreligious and drunken, but fond of a sing-along and a visit to the theatre. We often still use the word 'dustbin' to describe garbage receptacles, although, for the most part, it is a long time since they contained much coal dust.

These workers laboured in rather remote areas and had little contact with people, but chimney sweeps, who were essential ancillary workers in the burning of coal, moved among the general public, where, according to Mayhew, they were looked down on and 'treated with contumely', because of their filthy appearance and the offensive smell that they gave off.

FIGURE 29 John Thomson, The Temperance Sweep, from *Street Life in London*.

Mayhew's contemporary, the photographer John Thomson, included a portrait of a 'Flying Dustman' in his collection *Street Life in London*. Two men, in ragged clothes and with dust-blackened faces, stand by a horse and cart. They were part of Thomson's project to photograph the itinerant workers of London and, clearly, the dustmen were a familiar sight in the city. He also photographed a chimney sweep. Blackened with coal dust 'The Temperance Sweep' holds the tools of his trade: brushes and cloths, and looks away from the camera. An inadvertent graffito on the wall gives him a chalky halo and draws attention to his broken hat. Besides him, leaning insouciantly against the wall is a young boy (Thomson 1877).

Fighting pollution

Just as fogs were part of the long history of London, so were suggestions for ways to deal with air pollution. In 1661 John Evelyn railed against the smoky air and proposed some ways of alleviating the problem:

> And what is all this but that hellish and dismal Cloud of SEA-COAL? which is not only perpetually imminent over her head … but so universally mixed with the otherwise wholesome and excellent air, that her inhabitants breath nothing but an impure and thick mist accompanied with a fuliginous and filthy vapour, which renders them obnoxious to a thousand inconveniences, corrupting the lungs, and disordering the entire habits of their bodies, so that catarrahs, physicks, coughs and consumptions rage more in this one City than in the whole earth besides. (Evelyn 1661: 5)

'Sea coal' was simply coal that had been carried to the city by sea, and Evelyn was not the first person to identify it as the source of pollution in London. His proposed remedy was to remove all industry from the city and to plant a *cordon sanitaire* of trees around the town itself. Not only would noisome trades be expelled, but even the dead would have to be buried outside the city walls. For several hundred years after Evelyn, reformers struggled against those who, from vested interest or real belief, argued either that air pollution was an inevitable companion to economic success, or that smoky air was positively beneficial. Stephen Mosley notes that the nineteenth-century citizens of Manchester saw the five hundred smoking chimneys that blighted the air quality as proof of their status as the world's first truly industrial city.

> By the 1880s, a century's experience of living with smoking chimneys had given a cultural permanence to the notion that coal smoke denoted wealth. The correlation between smoke and prosperity had become deeply embedded in the culture of northern industrial society that many city dwellers did not often think to complain about the murky atmospheric conditions. (Mosley 2004: 69)

This reminds us that pollution is a cultural matter as much as a simply scientific issue. The great nineteenth-century industrial cities were a product of coal. Steam power made it economic, or even essential, to bring workers together into large factories. The belching chimneys were proof that hundreds of people worked in the building beneath them. But Victorian medicine also saw lung diseases, fevers and infections as being caused by miasmas, rather than by the polluted atmosphere. Stephen Halliday explains the notion of the miasma:

> This belief held that most, if not all, disease was caused by inhaling air that was infected through exposure to corrupting matter. Such matter might be rotting corpses, the exhalations of other people already infected, sewage, or even rotting vegetation. The 'miasmatic' explanation of the cause of disease figured prominently in the long debates among the people who were responsible for combating the cholera epidemics that afflicted Britain, and particularly London, between 1831 and 1866. (Halliday 2001: 469)

Halliday also reminds us that even the great sanitary reformer Edwin Chadwick, in his *Report on the Sanitary Condition of the Labouring Population of Great Britain*, published in 1842, argued for the improvement of house drainage to remove noxious smells from dwellings.

In keeping with the medical orthodoxy of the day, Chadwick saw foul air, rather than the water supply, as the source of infection. This was fetid air that had drawn-in corrupting matter from animals, corpses and rotting organic material. Some people argued that the coal-laden air, that contained large quantities of sulphur, acted as an antiseptic to the miasma. Over the next hundred years not only did medical science recognize water-borne pathogens as key to many diseases, but also clean air groups began to develop institutions that were designed to tackle the problem. In an article about the attempts to clean up the atmosphere in both Britain and the United States, David Stradling and Peter Thorsheim draw our attention to the fact that, as so often, coal pollution was also given metaphorical significance:

> As evidence of their shared Victorian ideals, the upper classes of both nations worried that smoke gravely harmed the lower classes, stunting their moral and physical development. In both nations, smoke became symbolic of a fear of the working class and the increasingly visible poor. (Stradling and Thorsheim 1999: 7)

Smog and art

While social reformers deplored the pollution of great cities, long before the 1950s some artists recognized the sublimity of buildings, railways, traffic and trees seen through a veil of fog. Turner's *Thames Above Waterloo Bridge,* which he painted in the 1830s, but left unfinished, shows the city enshrouded in smoke that is belching from factory chimneys. Claude Monet, too, when he had finished his series of paintings of Rouen cathedral, turned his attention to London and made a number of images of the Waterloo and Charing Cross bridges. Christine L. Corton tells us that the artist Benjamin Robert Haydon praised the fog and described it as 'a sublime canopy'. The American artist James McNeill Whistler created his famous 'nocturne' paintings. He used the word to describe works that showed night scenes or those in which the light was obscured and veiled the subject rather than revealing it clearly (Corton 2015). Alvin Langdon Coburn's photographs taken in the fog share this aesthetic delight in the softening of the outlines of buildings and the muting of colour. His study of a tree in Kensington Gardens taken in 1909 demonstrates his symbolist sensibility and the influence of Japanese prints, as the flat surface renders the tree into a balanced design. He showed his work at the Royal Photographic Society and collaborated with a number of European artists. He was a major figure in American pictorialism and, over time, his work moved increasingly towards abstraction.

Other European cities were not so badly affected by smoke and fog as British ones, because they tended to burn coke in enclosed stoves rather than bituminous coal in open fires. Nevertheless, there were plenty of thick fogs. Yale Joel, a *Life* magazine photographer, captured carefully composed scenes of Paris in a dense fog in 1948, as, in a freer style, did the French photographer Willy Ronis. The distinguished street photographer Fred Stein played with two kinds of depiction in his 1934 photograph which showed a couple almost effaced by a soft Paris fog while above them looms a harsh, geometric roof.

While London became famous for its smog, it was not, of course, the only British city to suffer in the days before the clean air acts took effect. Not for nothing was part of the Midlands known as 'the Black Country', but undoubtedly one of the most polluted places was Manchester. Travellers to the city write of a 'canopy' of thick smoke hanging over the whole place and note that artificial light was necessary even at noon.

In the year 2000 a memorial statue by the sculptor Paul Vandersleyen was unveiled in Engis, a village in Belgium. It recorded the fact that sixty people were killed by fog in the heavily industrialized Meuse Valley In 1930. A rigorous enquiry into these events convinced scientists that it was, indeed, the terrible air quality that led to the deaths: a conviction that did little to prevent further fatalities in other places. In 1948 in Donora, Pennsylvania, more than twenty people were killed in dense smog that emanated from the zinc and steel works that were major employers in the town. Decades after the event respiratory illnesses were common in the town. In 1950 President Harry Truman convened the first national air pollution conference and he explicitly cited the events at Donora as evidence that a clean air act was needed.

FIGURE 30 Fred Stein's photograph of a couple at a street corner in fog-bound Paris in 1934.

London and New York

The greatest pea-souper in the twentieth century was the London fog of 1952. Even today it is regarded as the worst air pollution in European history. From the fifth to the ninth of December a vile, yellow fog blanketed the city and mixed with the smoke of domestic and industrial fires to produce deadly smog. There are hundreds of accounts from Londoners about the way the smog affected their lives. It brought transport to a halt, made moving around the city very difficult and killed some 4,000 people in the five days that it lasted. When it was all over the estimate of the number of casualties who finally suffered was around 12,000 deaths with thousands of people rendered ill. Pollution had been taken for granted as being simply a part of urban living, and the organizations dedicated to cleaning up the atmosphere were largely ignored. Now there was a general feeling that something needed to be done. The romance of fog, its ability to render the city strange and charming was no longer as important as the reality of smog that swirled through the streets and into buildings and brought death and disease with it. The smog of 1952 was an event that could not go unremarked; photojournalists showed stalled traffic, traffic lights dimly gleaming through the murk, the sky lowering and black, and people with masks and handkerchiefs pressed to their faces. They revealed the way in which a once-dynamic and mobile city had slowed almost to a halt, with cars, buses and pedestrians all edging forward into an unknown, almost invisible space. Policemen led with powerful flashlights and were reminiscent of the 'linkboys' of the nineteenth century who helped people around the streets by carrying flaming torches. The newspapers were full of human interest stories, but as Peter Thorsheim has noted:

> Newspaper reports during the 1952 fog said virtually nothing about its possible health effects. Most articles focused instead on the visual aspect of fog. Some newspapers, celebrating the ability of dense fog to transform the appearance of London, published ghostlike photographs that depicted the fog-enveloped city as mysterious and sublime. (Thorsheim 2006: 157)

While the press was little interested in the science of the outbreak, reformers and politicians knew that something now had to be done. The great smog led directly to the Clean Air Act of 1956 that introduced smoke control areas and shifted domestic heating away from open hearth fires to cleaner coal, electricity and gas. It was followed by a more comprehensive act in 1968. The Great London smog still

FIGURE 31 In 1956 smog masks were issued to London policemen.

receives attention today, but may other cities suffered at the time. Both Glasgow and Manchester were periodically shrouded in dense smog, while in the United States Chicago, Pittsburgh and St. Louis were among the cities that had to deal with severe outbreaks.

In 1905, Dr Henry Antoine Des Voeux delivered a paper to the Public Health Congress in London. In it he drew attention to the presence in urban areas of something he called 'smokey fog', or 'smog'. This neologism proved to have a very long life. Originally it was applied to pollution brought about by coal, but smog is now usually used to describe atmospheres affected by the emissions of motor vehicles combined with bright sunshine to produce 'photochemical smog'. While Los Angeles is often regarded as the capital city of smog, in recent years Beijing, New Delhi, Paris and London have all topped the pollution count at one time or another. In 2016 the World Health Organization published a list of the top thirty polluted cities. Delhi could only come in at number twenty-one, while Beijing did not even appear. At number one was the city of Zabol in Iran. This list is contentious, and the statistics difficult to verify, but, regardless of the ranking, it does demonstrate the worldwide nature of the problem.

We have seen that the London fogs were treated by artists and photographers as picturesque and romantic, softening as they did the harsh outlines of buildings, and bestowing on ordinary streets a sense of mystery. The response to smog in more recent years has been very different. There has been an attempt to use visual material in a heuristic way to show the global reach and consequences of pollution. A leading example of this was the 2011 exhibition *Coal and Ice*. This was described in *The New Yorker* as

> a documentary exhibition encompassing work by thirty photographers around the world. It seeks to do something unprecedented: to chart the horrific grandeur of our effects on the planet, from the coal mines beneath our feet to the dwindling glaciers on our highest mountains. (Osnos 2011: 7)

If 1952 was a crucial year for London, fourteen years later New York was plunged into a smog pollution that led to changes in attitudes and the law. For three days the city suffered high levels of pollutants as well as being shrouded in smoke. The event was closely studied and led eventually to changes in the law. The smog, it was decided, was caused both by emissions from cars and by the industrial burning of coal. The research also indicated that most large American cities suffered high levels of pollution, although New York City did top the league. Medical investigations indicated that the smog of 1966 led to premature deaths, and to significant numbers of people suffering respiratory problems.

Smog in London was treated partly as an intolerable event, but also as an almost nostalgic reminder of times gone by with the return of a full-blown 'London Particular' in 1962. In New York a more rational spirit prevailed. It was clear that the methods for measuring air pollution, let alone ameliorating it, were inadequate

FIGURE 32 Walter Albertin captured this view of the Chrysler Building in dense fog in 1953.

and would need to be improved. Support for environmental groups grew, and there was a call for action to deal with air pollution. Within two years a series of air monitoring stations had been opened and legislation to improve air quality was put into place. In 1970 a national commitment to creating clean air led to the Clean Air Act and the foundation of the nationwide Environmental Protection Agency. London and New York are both still polluted cities, as is much of the world. Now diesel fuel and sheer volumes of traffic are at the heart of the pollution. In Los Angeles smog was a problem in the 1950s and 1960s. Together with London, the city has given its name to a particular kind of smog. 'London smog' is a mixture of natural fogs combined with the industrial burning of coal. 'Los Angeles smog', on the other hand, is a product of the extensive burning of petroleum in cars. Photochemical smog is produced when pollutants react in sunlight. Of course, these are ideal types and most places around the world have a mixture of several sources and types of smog. Imaginative schemes to reduce road traffic have been tried out in Paris while London has a congestion charge that imposes a tax on vehicles that want to enter the city. In India and China extremely high levels of smog are still quite usual. In Beijing photographers such as Alexander F. Yuan, Ed Jones and Fred Dufour have shown the city almost blanked out by the dense cloud of pollution that veils it. Despite the multiple causes of smog in China, from pollution by industry, vehicles and the generation of electricity, it is often reckoned that the chief cause of more than 300,000 premature deaths linked to air pollution is the burning of coal. The Chinese government is taking measures to try to ameliorate the problem, but coal is likely to remain the chief source of heating for some time to come. Beijing's assumed status as smog capital of the world is being challenged by Delhi which, according to the World Health Organisation, is one of the most polluted cities in the world, and has staggeringly high levels of pollution in the winter months, much of it from the burning of coal. Visitors to the city, *The Telegraph* reported, are so impressed by the atmosphere that they take selfies in the smog (Oliver Smith, *Telegraph*, 9 November 2016). Delhi's smog is the product a mixture of coal burning and petrochemical sources. This is a particularly toxic mix. Recent research seems to indicate that coal-burning power plants and diesel vehicles may produce pollution that is more harmful to human health than other types of pollution. All this has led to a great interest by artists and environmental activists to act together in order to draw attention to the seriousness of the problem. The BREATHE exhibition is a collaboration between the World Health Organisation and Carbon Arts. In its catalogue to the exhibition it says that:

> Globally, there is a growing recognition that addressing air pollution, climate change, and related health and environment issues, is not simply a scientific or technical problem, but a cultural one. To a large extent, we already possess the technology and best practice policy examples needed to ensure the well-being of people and make our future sustainable, but these aren't always enough to drive change. At heart, responding effectively means confronting our personal identity, worldviews, and sense of loss or adventure experienced from any sort of change. (Denfield and Newcombe, 2015 coclimate.com)

Smog has moved away from being regarded as simply a scientific, technological and medical problem, to taking centre stage on a global movement to convince us of the fragility of the planet and the need, at the international, national, local and individual levels, to reduce our use of dirty sources of power. The struggles in London or New York to create clean air acts are magnified in a worldwide campaign to lower our carbon emissions. The very air we breathe has become an issue for cultural politics as well as for legislators. Changes in climate technology have a huge influence on the nature of human societies, and many cultural and art groups are working to make climate change more resonant with the population as a whole and explore the nature of our own subjectivities in response to the scientific study of climate and pollution.

At least at the formal level it seems that governments are responding to the problems caused by burning coal. In 2016 the British government announced that the last coal-burning power station in the country would be closed within a decade. They would be replaced with gas plants and are expected to reduce emissions significantly. When running for president, Emmanuel Macron said that he would close all coal power plants in France by 2022 and ban oil and gas exploration in French territorial waters. Coal makes up only about 3 per cent of France's power generation, which is largely run on nuclear power. Canada, which has already significantly reduced its use of coal to about 7 per cent of its energy generation, announced a total phaseout of the fuel by 2030. Germany has pledged to phase out half the coal-fired plants by 2030 and the rest some twenty years later. Finland will abandon the use of coal by 2030 and currently imports coal to fire some 12 per cent of its power stations. The Netherlands and Austria have similar ambitions. An agreement within the United Nations Framework Convention on Climate Change, known as the Paris Climate Accord due to come into force in 2020, aims to hold the increase in global average temperatures to below 2 per cent. Each country is free to make its own plans in order to achieve this aim, but, clearly, the burning of coal would have to be severely curtailed on a global level if the central aim is to be achieved.

In 2016 Zhang Lei won first prize in the contemporary issues section of the World Press Photo competition for his photograph *Haze in China*. It shows a cloud of smoke hanging in Tianjin in northeast China. 'Haze' is an interesting word for the picture, taken from a high vantage point, showing a ranked series of tower blocks reflecting stray gleams of light while a dense blue fog blanks out most of the city. As with the photographs of London and New York, the obscuring of this everyday scene lends it a touch of beauty.

Pollution

Coal plants are a major source of carbon dioxide emissions, but in his comprehensive survey of the diseases caused by the burning of coal, Alan H. Lockwood says that the public are unaware of the multiple ways in which the pollutants from coal damage our health. He notes that the different types of coal have different effects, but then goes on to look in detail at the pollutants spewed into the atmosphere when coal is burnt (Lockwood 2012).

These are arsenic, oxides of sulphur and oxides of nitrogen, which produce ground-level ozone, a major component of smog. Also released are particulate matter which is a key component of urban pollution, but also affects those people, now largely in developing countries, who burn coal on open fires. Coal also puts mercury, an intensely toxic substance, into the air.

Add to this the many diseases that miners contract when digging it out of the ground and the case against coal becomes more compelling. In addition, burning coal produces ash. This ash has to be disposed of, and around the world more than 500 million tons of ash are stored in large tanks or 'impoundments'. In 2008 an impoundment ruptured and some 5.4 million cubic yards of ash poured down onto the town of Kingston, Tennessee. It covered roads and railway lines, and polluted the rivers as quantities of heavy metals were released into the water supply. It took eight years for the ecosystem return to pre-spill levels.

Despite these many problems, some people still believe that coal has a viable future if we can burn 'clean coal'. Coal's status as the dirtiest of all fossil fuels means that most countries have pledged to stop using it within a defined time period. 'Clean coal' refers to a cluster of technologies that attempt to remove the emissions from coal. Washing coal to take away the dross that clings to it, or scrubbing it to remove sulphur dioxide, can achieve this – although both have problems with the disposal of the residues. The most important technique in order to produce 'clean coal' is that of 'carbon capture and storage'. Essentially this involves collecting the emissions from burning coal and storing them below ground. It is claimed this would dramatically reduce the amount of carbon dioxide in the atmosphere, by as much as 90 per cent. The economics of the technology, however, also needs to be taken into account. Carbon capture and storage (CCS) is expensive even when built into new power stations. Retrofitting them to existing stations is difficult and even more expensive. The Carbon Capture and Storage Association declares that CCS is a key tool in tackling climate change, providing energy security, creating jobs and ensuring economic prosperity. Some critics argue that the technologies are relatively untested and that, in any case, employing them might encourage the continued or enhanced burning of coal as this would, apparently, have fewer deleterious consequences. They are anxious, as are many governments, to see the end of coal all together. But we are a long way from that. Indeed, in October 2017 the American Environmental Protection Agency announced that it would repeal President Obama's Clean Power Plan that was designed to limit greenhouse gas emissions. Environmental groups are fighting this repeal.

The widespread and global political action over air pollution reminds us that fogs and smogs have never been pollutants that could be simply dealt with. Rather, they have been caught up in cultural and political debates. There was a long struggle in Britain to put clean air acts on the statute book, despite the choking fogs, the acid rains, the marred buildings and the choked lungs. Culturally, fogs were seen as natural phenomena, and it was once thought that they originated in the countryside and drifted into the towns and cities. There, they transformed the appearance of things and made the ordinary mysterious. A blazing fire summoned up notions of happy family life as everyone gathered around the hearth. Even when it was clear that dense fogs were the product of domestic and factory fires, there was pride in the fact that the output of industry could be read in the density of the coal-laden air.

At the end of the nineteenth century the artist and antismoke leader William Blake Richmond echoed the notion that environmental changes could mitigate class conflict. Once the air was free from smoke, he asserted, poor people would 'lose their sullen looks and become more bright and cheerful'. William Booth, the founder of the Salvation army, shared this perspective and blamed 'the foul and poisoned air' of large cities for contributing to drunkenness. Men drank, he argued, to compensate for the inadequate supply of vitalizing oxygen and ozone in the air of cities. (Thornsheim 2006: 119)

Cleaner air, however, did not lead to a diminution in class conflict, and in the next chapter I want to look at some famous strikes in the coal industry that were of great importance and still resonate in popular culture.

5 STRIKES AND CONFLICT

One of the things that everyone knows about miners is that they have always been militant fighters in class struggles, and could often be found at the heart of working class industrial action. In part their power came from the fact that coal was the basic commodity on which all other industrial production depended, so that, in the right circumstances, miners could wield considerable influence. On the other hand, coal could be stored for indefinite periods, so a steady supply might still be available even while the miners were on strike. Perhaps more importantly their strength derived from the collective nature of the work and the tight-knit supportive communities that grew up around mining.

Strikes were usually called as part of the struggle to create or join trade unions, to improve safety in a mine or to increase wages. Because of the ways in which miners were paid, negotiations on wages were often prolonged and complex. Even what would appear to be straightforward – the question of how much coal a man had cut – was a matter for negotiation, as managers claimed that there was too much dust and small coal in a wagon, or weighed it as much less than the miner had calculated. In the 1930s in South Wales, B. L. Coombes described the intricate negotiations that would set a price for the various tasks involved in cutting a ton of coal:

> Now there was a better supply of air in the workings the management wanted the price list settled without delay. There were nightly committee meetings and general meetings to settle what it was worth to stand posts, cut the bottom, rip the top, work in water, drive airways, fix cogs, notch double timber, and a score of other jobs. Most of these items were settled without much delay, but the greatest discussion was over the cutting price of coal. (Coombes 1939: 83)

Coombes was an English immigrant to South Wales, where he worked in coal mines for more than forty years. He became one of the best-known working class writers of the 1930s when his autobiographical book *These Poor Hands* was published. It has become one of the classic works on mining, with its description of work in a colliery, the introduction of mechanical cutting, the processes of wage bargaining, accidents, deaths, strikes, lockouts and unemployment. He also describes the landscape, the housing stock and the everyday life of the village.

The negotiations he describes have always gone on. They are one reason why mining cannot be directly compared with manufacturing, where the conditions of work are relatively transparent and understood by all sides. But just as the nature of getting coal was problematical, so was (and is) the industry as a whole. Even when coal was the unrivalled fuel that underpinned the entire economy, the world market for it was always fluctuating. The demand for coal was driven by the fortunes of economies as a whole. It rose in times of war and high industrial output, but declined in economic downturns. A coal mine had to be kept open at all times, and the ease with which new seams could be found and mined was uncertain. In consequence, it was easy enough to lose money by running a small mine. One result of this uncertainty was that large conglomerates became the dominant form of ownership in most countries. These were able to use their oligarchic power to suppress wages and attempt to limit the activities of trade unions.
John Williams has argued that:

All the variations in the pace and direction of development, all the swings from relative quiescence to tumultuous conflict, can be seen as ultimately stemming from the attempts to adjust the level and nature of the supply of coal for the changes in demand for it. (Williams 1980: 155)

Adjusting those levels has rarely been a simple matter, and, despite the power of mine owners and managers, strikes have been common in every country where coal is mined. Take, for instance, the 1949 Australian coal strike, notable for the fact it was the first time in that country that troops had been used to break a strike. A year earlier French miners had struck for some two months and seriously damaged the economy. In the vital period of post-war recovery it has been estimated that some 5 million tons of coal were lost and the state was deprived of much needed revenue. Decades earlier the Cape Breton strike was a radical fight for both unionization and improved wages. The militant clashes went on for four years, and the coal and steel owners, who were castigated by a royal commission, became bankrupt in 1926. New Zealand's Great Strike of 1913/1914 was triggered by the dismissal of men seen as radical and therefore undesirable. Some strikes have also been directly political. The Soviet miners' strike of 1991 helped to change the course of Russian history as the miners supported Boris Yeltsin, and one of their chief demands was the resignation of President Gorbachev. However, this strike marked the apex of working class action in that country and later the miners saw a decline in their living standards and the gradual erosion of their industry. No strike stands as a single event. Each brings with it a particular set of demands and involves political struggle and community action. They can also involve an entire place and community. Take the 1932 strike in Belgium. At the heart of it was the Borinage region where Van Gogh had lived among the miners and painted them in a series of melancholy images. Here, too, the Dutch-born artist Henry Luyten recorded the strike in his huge painting called 'The Strike'. The struggle also inspired one of the most famous documentary films of the 1930s: Joris Ivens and Henri Storck's *Misère au Borinage*, which looked at the woeful condition of the miners, at evictions, the strike and the relationship of this particular struggle to capitalism as a whole. The film ends with a call for the death of capitalism and the rise of the proletariat.

The British miners' strike of 1984/1985

Many strikes might be discussed, then, each of them of great interest and national importance. I have chosen to write about a few, less for the political positions they exemplify, than for the ways in which they were reported and entered public consciousness. They were represented in film, photography, television, song and literature. Indeed, as we shall see, in one instance a strike was brought back from the amnesia of history by a feature film.

On the 23rd of April 2004 John Lichfield reported in *The Independent* newspaper that 'the French coal miner, a powerful symbol of social revolt and industrial strength for more than a century, passed into extinction'. The headline to the piece was 'France Ends Coal Mining with Tears But Not a Single Protest'. France was the first of the large industrial nations to abandon the mining of coal in favour of increasingly cheap imports. At that time Britain still had sixteen pits left and 4,000 miners, although that was a dramatic fall from the 170 pits and 180,000 miners that existed in the mid-1980s. The rundown of the industry in Britain was accompanied by plenty of tears and a great deal of strife. Indeed, the miners' strike of 1984/1985 was one of the most prolonged and ferocious in industrial history, and its effects were felt in the society as a whole and not just in coal-mining communities.

The aim of the miners was to oppose the National Coal Board's (NCB) decision to phase out deep coal mining where it was deemed to be uneconomic. This, in turn, was partly in response to the government's decision to sharply reduce subsidies for the industry. The miners had been successful in two earlier hard-fought strikes. That of 1970 was a seven-week stoppage that came about directly over pay levels. Miners were not badly remunerated, but had slipped down the pay league and all workers were feeling the pinch at a time of high inflation. This strike was notable because it was the first official miners' strike since 1926, although there had been plenty of stoppages at the local level. It was also significant because of the tactics used by the miners, especially the so-called flying pickets by means of which miners picketed workers not directly affected by the strike. They eventually won and received a significant pay rise, a decision that was supported by a Committee of Enquiry led by Lord Wilberforce. However, by 1974, the miners' pay had fallen from the level recommended by that enquiry, the British economy was in an even worse state and inflation was at record levels, not least because of a huge rise in the price of oil on world markets. The second official strike took place in that year and lasted for a month. The government tried to protect electricity supplies by rationing power to three days a week. In the middle of this highly unpopular and unsustainable position, the Conservative prime minister of the day, Edward Heath, called a general election asking the British public, 'Who governs Britain?' Whatever the answer to this question, he narrowly lost the election and it seemed that the workers could claim a significant victory. Under the new minority Labour government the miners got their pay increase and the strike was ended.

Nevertheless, despite the presence of a Labour government, a slow but steady pit closure programme went on across the country. By the mid-1980s, a new Conservative government was in place with a radical neo-liberal economic policy, and a determination to crush the power of the trade unions, in the vanguard of which was the National Union of Mineworkers (NUM). A series of changes to the law made it more difficult for workers to take direct action and the Government, which owned the mines, carefully prepared for an expected strike. It seemed to many people in the industry that not only clearly uneconomic pits were being closed, but that an undeclared strategy to run down the entire industry was in place. The definition of 'uneconomic' in this context became highly problematized, and there was a vigorous debate about what the future of coal might be.

While this was going on, the NUM, led by Arthur Scargill, was determined to save jobs by repeating the actions that had been so successful in the 1970s. These included restricting the supply of coal on which electricity supplies depended, as well as using the formerly successful weapons of picketing and mass rallies. Smarting from the earlier defeat, and determined to change the basis of the British economy, the government of Margaret Thatcher built up coal supplies, supported mining areas which did not take part in the strike and mobilized the police, creating a *de facto* national police force, and drawing thousands of officers from around the country into the coal-mining areas. As no national ballot of miners had taken place by the NUM, the Government went to court and the miners' action was declared illegal in September 1984. Despite this it became one of the bitterest strikes in British history with regular battles between pickets and the police. These confrontations became almost nightly television spectacles. The strike was generally understood to stand for more than the future of the mining industry. It was a fight between competing versions of how the economy should be run, but it also was an overt and dramatic example of class struggle in action. Prime Minister Margaret Thatcher famously declared that while the Argentinians had been an external enemy in the struggle for the Falkland Islands, the miners were 'the enemy within'. Using the huge resources of the state the Government finally defeated the miners in March of 1985, when they could hold out no longer and were forced back to work.

It was clear that if the strike were lost a radical realignment of the economy would take place with traditional, heavy industries likely to be less important than they were formerly. In December 1984 the coal industry was privatized, dramatically reduced in size and phased out over the next twenty years.

The industrial action was, perhaps surprisingly, supported by a large portion of the public at least to the extent that they contributed cash and food to the many volunteer collectors in towns and cities far from the mining villages. Here the image of miners as having appalling jobs and labouring in the dark far from the comforts of modern life probably had an effect, as did the fact that this was not a strike about wages, but was provoked simply by a desire to hold on to these unenviable jobs.

The strikers were also directly supported by filmmakers, artists, writers and photographers. One important cultural output was the *Miners' Campaign Tapes*, a project that began at the beginning of the strike. The films were made by a number of independent video companies and community activists. Chris Reeves, a member of Platform Films, has described their motivation:

> We saw the strike as a response to a deliberate planned attack on the working-class movement by an authoritarian Conservative Government, part of whose aim was to end the leading role of the NUM in that movement. (Reeves n.d.)

They felt that the balance needed to be redressed because the popular British press had, in their view, become one of the arms of the state and carried out a sustained attack on the miners, especially on Scargill and other leaders of the NUM. Throughout the strike it seemed that there were nightly scenes of violence as pickets fought with police at power stations or coking plants. These are frankly partisan tapes, and the titles of some of them that have now been released on DVD are interesting as they cover most of the issues that became central to debates about the strike.

Not Just Tea and Sandwiches was concerned with the role of women in the strike who often took part in picketing and rejected the notion that their function was to give domestic support to the men. The emergence of women as active players in the strike was one of the features much commented on at the time. Women who had never taken part in public life spoke at large meetings and rallies; they were interviewed extensively and some later published books about their experience and the fact that their lives and that of their communities would never be the same again.

The Coal Board's Butchery argued that the Government's claim that they intended to close only twenty pits was absurd and that they would work with the NCB to create the conditions under which increasing numbers of pits would become clearly uneconomic. This would be achieved by starving pits of finance while investing heavily in nuclear energy.

Solidarity – this film stressed the communal action and the working class solidarity of the miners. It had to deal with the fact that an important segment of the mining community, those who worked in Nottingham, refused to take part in the strike. The conventional media stressed the way in which the police had to provide escorts for these men in order to get them to the colliery.

Straight Speaking – this phrase was a quote from the head of the Coal Board, Ian Macgregor, who had been branded by Arthur Scargill 'the American butcher of British industry', although he was, in fact, Scottish by birth. However, he had already headed up British Steel, where he implemented major cuts and redundancies.

The Lie Machine concentrated on the media, especially the tabloid press, and the ways in which it supported the government and the NCB and made little attempt at dispassionate reportage.

Only Doing Their Job was concerned with the actions of the police; with the extraordinary numbers who were drafted into the struggle. Miners reported being stopped on motorways and made to return home despite having committed no crime or misdemeanour. They alleged extreme brutality in the mêlées at power plants and believed that the police had been militarized and were being used by the government not to keep the peace but to foment trouble.

The themes of all these films were much discussed at the time, and some of the events of the strike are still live issues in public discourse. Perhaps the best example of this is the events surrounding what became known as the *Battle of Orgreave*. This was a crucial struggle in the progress of the strike. On 18 June 1984 police and pickets clashed at a coking plant in Orgreave, South Yorkshire. It was very widely reported and in the years since the actions of the police and of some media companies have been criticized. After the official documents on Orgreave were released in 2017, David Conn, writing in *The Guardian* newspaper, concluded that 'What happened at Orgreave was not simply the most violent police behaviour ever seen in a modern industrial dispute, but the culmination of a concerted political campaign to diminish the strength of trade unions' (Conn 2017: 26).

In line with this political position, in June 2001 the artist Jeremy Deller re-staged the battle. This action was an art event, a political statement, a rallying of popular memory, and direct recollection of significant events. In his book *The English Civil War Part II: Personal Accounts of the 1984–85 Miners' Strike*, he presents a series of accounts of the events of the struggle, together with material on his re-enactment. It is supportive of the miners' case and action, but published some seventeen years after the event can now be read as a piece of history (Deller 2002). It is a history grounded in the memories and in the voices of those who took part in the strike, as opposed to an academic analysis, and its use of photographs points this up. They are not chosen for their formal or aesthetic qualities but for how much of the action they reveal. They are photographs of the struggle from a number of angles, including police photos taken from behind their lines.

In one of the accounts Howard Giles recalls one of the lighter moments that happened among the mêlée: 'In one of those surreal moments that no-one would believe in fiction, let alone reality, an ice cream van – Rock on Tommy – continued to sell ice cream while it was completely surrounded by advancing police' (Deller 2002: 030). A photograph by Neville Pyne shows Rock-on Tommy selling his wares

FIGURE 33 One of John Sturrock's many photographs of the strike. This was taken at Orgreave in 1984.

surrounded by running miners and mounted policemen. The van is also present in the re-enacted battle. Bizarre events of this kind were rare, however, and the personal accounts of the strike are consonant with those that appeared at the time. One interesting feature is the role of women in the struggle, exemplified in this book by Stephanie Gregory, who recounts her growing involvement in the strike. Typically her entry into the public realm, organizing, speaking in public, and attending meetings, changed her life. Partly this was an example of a transformation in women's domestic role as miners' wives who were traditionally expected to concern themselves with cooking, cleaning and child rearing. She says that:

> The biggest change for me personally was the confidence that I gained in communicating with official bodies, managers of the Benefit Agency, Trade Union Leaders, social workers, reporters, Police Officers at different levels in the hierarchy etc. There weren't many doors that we didn't knock on and in the end I wasn't afraid to talk to anyone. (Deller 2002: 038)

Deller's restaging of the Orgreave struggle was planned with extraordinary precision. It used battle re-enactment groups, who were used to staging historical battles, and was orchestrated by a leading figure in that world, Howard Giles. Deller and Giles diligently researched the events of the day and studied oral accounts, newspaper reports, TV programmes and film footage. It was important to show the ad hoc confusion of the day as well as reveal the careful preparation of the police. This is perhaps the most ambitious project to have been created by the strike. Its aim was, one assumes, to return to the events of the day free from the accretions of media reportage or special pleading from interested groups.

The strike and the mining communities were also recorded in collections of photographs made in 1984–1985. Keith Pattison published a book of his pictures of the strike under the title *No Redemption* in 2010. He had spent months in the mining village of Easington in County Durham picturing everyday life. In this extraordinary time ordinary existence included eating at soup kitchens, standing on picket lines or gathering sea coal from the shore. Here mothers with their children are casually chatting while a line of police in riot gear stand behind them. The public space of the village had been occupied by the police forces, but the domestic spaces remained much as they always were (Pattison 2010).

In the same mode was a collection by a young German photographer Michael Kerstgens, *Coal Not Dole: The Miners' Strike of 1984/85*. He worked in South Wales and in Yorkshire and diligently photographed in the workingmen's clubs, the streets and the picket lines. His status as an outsider shows in some of the images, few of which would be defined as authoritative statements about the real nature of things. But this blurring of cultural focus is not confined to photographers from outside Britain; it is present in almost all the documentary work of the time. We may see this as a lack of intimacy with mining communities, but it may also be because the documentary form itself was becoming increasingly questioned.

Documentary photography was important to the miners' struggle not least because it humanized the miners and their communities and liberated them from the demonic accounts of much of the mainstream media. Reportage, oral accounts and photographs all helped to counter the dominant narrative that supported the notion of the miners as the 'enemy within'. There was little dispassionate output during the strike. Cultural workers took sides in the struggle as surely as did any other group. Photography was powerful not least because of the plethora of magazines, news sheets and journals that existed at the time. Many of these were well established and supported the documentary movement, but some were created for the strike itself. Tony Harcup has written about the alternative press and its role in the conflict. In a case study of the tiny *Leeds Other Paper* (LOP), an independent magazine, run by a collective for twenty years from 1974, Harcup writes:

> The paper's contacts and coverage took time to build up, but from June 1984 onwards the numbers of miners and miners' wives directly quoted in the paper increased markedly, and reportage from picket lines, soup kitchens and pit villages became a staple of its coverage. These were overwhelmingly 'ordinary' rank and file strikers rather than full time trade union officials. The name of union leader Arthur Scargill was scarcely mentioned in a whole year of LOP coverage. (Harcup 2011: 27)

In addition to community journals and magazines there were also independent photography magazines and photo agencies that were integral to the work that went on in support of the miners. The clashes at picket lines and power stations were regularly recorded with several diligent photographers in the vanguard. For example, John Sturrock was always to be found in the heart of the action. He recorded the many battles from street level for publication in the *Socialist Worker*. John Harris, who also followed the action and photographed at Orgreave, took the most reproduced photograph of the strike. It showed another photographer, Lesley Boulton, about to be hit by a truncheon-wielding mounted policeman who is looming over her. It was used to create a powerful poster that was widely distributed.

Martin Shakeshaft also photographed at Orgreave and Chris Killip recorded the village of Easington as a militarized place with police around every corner, while Imogen Young focused on South Wales. These photographs showed both the militant struggles with police and the nature of life in pit villages with communal

FIGURE 34 John Harris' celebrated shot of a mounted policeman attacking Lesley Boulton from the Miners Women's Support Group.

eating places, food distribution points, coal picking from tips and chatting in the street. Several people concentrated on the role of women in the strike. Raissa Page, for example, covered the activities of the Women's Action Group and the Women's Support Groups.

In 1985 Artworker Books produced a collection of photographs designed to support the miners. It includes all the major events, including the march back to work. Many of the photographers who covered the strike are represented in it. One of Chris Killip's Easington photos is included, showing a man lounging in an open doorway with a mug of tea in his hand while just around the corner is a phalanx of police in helmets and carrying riot shields. John Sturrock was present at power station pickets and demonstrations, while Raissa Page showed us family life and the work of the women's action group. This little book is probably the most lively and coherent single collection of photographs of the strike. Cleverly edited it balances the dramatic clashes with police together with the continuities of everyday life in the mining communities (Huddle et al. 1985).

Women's groups

The important role of women in the strike was also much written about and discussed. Women from mining families were visible on the picket lines, in the villages, on television and in print. They also contributed extensively to the oral histories and life stories that were being made at the time. Many people saw this vital contribution to the struggle as unique to the 1984 strike. However, in her

FIGURE 35 Roger Tiley, *Glynis and Mary, Members of Maerdy Wives Support Group, Maerdy, Rhondda Fach, 1985*. From The Valleys Archive.

article 'Women of the British Coalfields on strike in 1926 and 1984' Jaclyn Gier-Viskovatoff traces similarities between that strike and the famous general strike of 1926. Jointly produced with the photographer Abigail Porter, the article tells us that some popular forms of protest survived from the nineteenth century and 'certain elements have remained part of the repertoire of vernacular culture in coalfield society even to the present day' (Gier-Viskovatoff and Porter 1998). These include taunting blackleg miners with women's clothing hung on sticks to question their masculinity, confronting strikebreakers as they tried to go to work and organizing and running soup kitchens and other community resources. This feminine specialization on frustrating strikebreakers would seem not to be either new or confined to Britain. Mother Jones, the celebrated American supporter of miners and working people generally, wrote that 'women kept continual watch of the mines to see that the company did not bring in scabs. Every day women with brooms or mops in one hand and babies in the other arm wrapped in little blankets, went to the mines and watched that no one went in. And all night long they kept watch' (Jones 1925: 17).

Priscilla Long goes further, seeing women in America as central to the combativeness of mining communities:

The aggressive participation of wives and daughters was a root cause of the particular militancy of the mining communities. The women's work in the coal camps, their perceptions, and their consciousness and the activism that grew out of it are vital to the history of coal mining in the United States. (Long 1989: xxiv)

The role of women in the strike was not only discussed at the time, but also gave rise to a number of oral history accounts and memoirs by women. The Welsh Campaign for Civil and Political Liberties, together with the NUM, produced a book in 1985, described in the Preface as a 'special book' and as 'the first chapter of a people's history of the 1984/85 strike'. Here are accounts of women's experiences of being on the picket line. One woman described the surprise of the police to see them there. 'The police were surprised because we were women. They didn't know where to put their hands, because when they touched us we gave them abuse. They didn't know how to react'. But another noted that 'The police treated us all with contempt. They didn't care if it was a woman or a man they were dealing with'. Organized as the Women's Support Groups, they also carried out the more traditional role of cooking in communal kitchens and distributing food. Several women describe their experiences in the strike, which were outside their normal day-to-day existence, as 'life changing'. One respondent in Jill Miller's book *You Can't Kill the Spirit* says:

I wouldn't have wanted to miss this strike. It's taught me more in a year than I managed to learn in the rest of my life. It's made me aware of how bright and resourceful we are. Maybe the part we women played doesn't seem much to write home about, but believe me it gave us all a hell of a shake up and a boost to our lives. (Miller 1986: 51)

These are reminders that major strikes are more than just ways of fighting for changes at the workplace. They influence the values of the communities within which they take place, and sometimes resonate in the dominant national culture as well.

Retrospective accounts

In 1985 at the end of the strike Ken Loach's film *Which Side Are You On?* was released. There is no narrative voice, but a beautifully edited collage of 'songs poems and experiences of the miners' strike 1984'. In living rooms and social clubs people tell the story of the strike and analyse its causes and progress. They also recite their own poems and listen to stirring musical ballads and campaign songs. There is a good deal of live action footage at picket lines and in villages. One constant theme reiterated by miners in the film is that the struggle is for the next generation, so that they can continue to find work in the pits and to support the life of the local communities.

In a more allusive way Tony Harrison's long poem 'V.' was written just after the miners' return to work and infused with the bitter memory of the strike. It received a great deal of attention. At least it did when a Richard Eyre film of it was shown on television. The programme was criticized for its prolific use of obscene language. *The Daily Mail* called it a 'torrent of filth', and questions were raised in Parliament. In the poem the letter 'V' stands for several things: for victory, for a derisive sign and for 'versus'. It echoes Gray's 'Elegy Written in a Country Churchyard' and is set in a Leeds cemetery where Harrison's parents are buried. It is a cemetery that sits above a disused coal mine; a mysterious subterranean space into which it will one day sink; it has also been vandalized by football supporters. Harrison conducts a fiery debate with the putative graffiti writer, both upbraiding him for his yobbish behaviour, and sympathizing with him for his empty existence, for being denied work and for having no authentic source of satisfaction in his life. Harrison comes from the same kind of community as the football hooligan and explores some of the ways in which class bears on life chances and lived experience. He might himself, with different childhood experiences, have ended up as one of these boys (Harrison 1985).

One of the most curious films to address the miners' strike directly is the 2014 movie *Pride*. Written by Stephen Beresford and directed by Matthew Warchus, *Pride* won the Best British Independent Film at the 2014 British Independent Film Awards and was nominated for three BAFTA awards. In the summer of 1984 with the miners' strike in full swing, a group of gay and lesbian participants in the London Gay Pride march decide to support the striking miners and, like many groups, raise money to support the families. They become part of the Lesbians and

Gays Support the Miners group (LGSM). Their benevolent act is, however, turned down by the National Union of Mineworkers who are not happy at being associated with gay people. This might well have ended their activity, but they decide to bypass the Union, choose a plausible community in the Welsh valleys and make their donation directly to the striking families. They arrive in the darkened village of Onllwyn and present themselves to the bemused villagers. Both sides have preconceptions about the other, and the film is very good on the attempts to forge bonds. One of the villagers tips off the tabloid press, and the miners face catcalls from the police as they picket. The gay group are deeply experienced in police harassment and give the community good legal advice when they are arbitrarily arrested. The NUM still urges the miners to cut their ties with the LGSM. Having been called 'perverts' by the tabloid press, they run a 'Perverts and Pits' festival in London which some of the villagers attend with great enthusiasm. However, the Union manages to run a meeting that excludes the gay group from taking any further part in the action. At the Gay Pride march, 1985, *our* group has been asked to walk at the back with the fringe groups because the organizers want to suppress any potential political action. Then they are moved to the front because not only are there a handful of villagers in their ranks, but hundreds of miners led by the NUM have arrived to march in solidarity. This simple story does not do justice to some of the political and cultural threads of the film. At the end, one of the miners has come out as gay and the role of the women's group Lesbians Against Pit Closure has been examined. Two very different disadvantaged groups come together and, from the distance of the twenty first century, attitudes to both the miners and the lesbian and gay community have been explored. Essentially, though, it is a feel-good movie in which the 'otherness' of the LGSM and the miners is buried in a collective experience.

In 2017 The Whitworth Gallery in Manchester showed an exhibition of work about the miners' strike. Curated by Craig Oldham, *In Loving Memory of Work: A Visual Record of the UK Miners' Strike* was based on Oldham's book with the same title. It is a fascinating collection of texts, photographs, posters, record covers, graffiti and badges, together with articles and personal accounts of the struggle. It is a reminder of the power of what might be regarded as mere ephemera to create and sustain emotions of solidarity and resilience. There is an immediacy in many of the artefacts, some of which are no more than scrawls on a wall that nevertheless convey the urgency, the importance and sometimes the humour of the events taking place. It also shows the photographs of a young miner, Peter Winnard, who took his camera along to the picket lines.

> With it, Winnard captured the camaraderie and community shown throughout the dispute; the wit of the placards and the posters; moments striving for solidarity with other trade union workers; and the now historic gathering of thousands of women from mining communities in Barnsley, 1984, to start the women's movement against pit closures. (Oldham 2015: 34)

Most of the photographers who covered the strike were professionals, but Winnard came closest to emulating the mission of the workers' photography projects of the 1930s. In a piece called 'A Discussion on Dissent', Rick Poyner was asked how he saw the graphic images of the strike, many of which were produced hastily and with no thought for their longevity. He said:

> These images are important and have value because they are the spontaneous, authentic and urgent expression of political ideas and demands and of the motivations of the strikers. (Oldham 2015: 11)

If this work ought to be preserved we may ask what connections there are between events that took place more than thirty years ago and our present position. In a magazine interview Craig Oldham was asked about any future political projects he might undertake:

> There certainly are the similarities between 1984–85 and now, but the time is different. I remain politically involved, and to a degree get involved when I can, but I don't see the miners' fight as one that is over, but rather ongoing, and I'm still fighting that. (Oldham 2016: n.p.n.)

This notion that the strike is still with us and still potent was given further impetus when, in 2016, the Home Secretary, Amber Rudd, announced that there would be no enquiry into the events at Orgreave, despite repeated calls from miners and libertarian organizations for a review. The fact that this was a widely reported and frequently discussed decision demonstrates that the miners' strike is still an important event that has helped shape the British political imagination. But, for all the continued interest in the strike, and in the cultural artefacts that it produced, the former mining communities have never recovered from the loss of their industry. They are complex places where many people are making new and fulfilling lives, but they have also suffered from economic stagnation and social dislocation. In an interview David Douglass described the post-strike mining villages in this way:

> All of it has changed and run down, it's riddled with poverty, most people on benefits, or trying to work on the black market, anti-social crime, heroin addiction, anti-social violence. Which we could have predicted: if you take the pit out of the pit community what is it there for? All of the things that people considered to be meaningful in their lives, their history that they'd literally heard from grandfathers, their parents, dads, grandfathers and great grandfathers all working in the pits, they lost that. (Deller 2002: 20)

At the level of representation, the strike was communicated, and made comprehensible, through photojournalism and documentary photography. It was, indeed, one of the last major events in the UK to be articulated in a comprehensive way through still photography. The infrastructure of radical picture agencies, magazines and journals no longer exists. Digital technology has changed the nature of things. At the time there were no tweets or selfies, and one could not follow the miners on Facebook. But the next strike I look at impacted on the world through a careful documentary that followed events over a long time.

In Harlan County

Barbara Kopple's film *Harlan County, USA*, won an Oscar in 1976 for best documentary. It is the story of a strike against the Duke Power Company by miners at the Brookside Mine in South East Kentucky in 1974. They wanted a new contract that would improve safety, increase wages and establish a new code of labour practices. Kopple and her film crew spent a great deal of time in the town and managed to capture a range of activities from picketing to family life. They interviewed miners suffering from pulmonary diseases and listened to the general talk. After an opening sequence of scenes of underground work, the film cuts to a couple on a porch who tell us something of the nature of life in the place. There is no authoritative narration, so the film creates its complex story through the actions of the people in the community and from meetings held with representatives of the Duke Power Company. A combination of live footage, archive material and plaintive music, including Florence Reece singing 'Which Side Are You On?' – her famous question to striking miners – drives the film. Kopple judiciously uses archive footage, for example of John Lewis, the miners' leader, who was President of the United Mineworkers of America for forty years from 1920, picketing a mine.

The strike continued for almost a year. While the President of Duke Power claims that they have tried to settle the strike, and will meet many of the demands, he will not tolerate a union at the mine. The miners think that without a trade union any settlement will be hard to maintain. They also point out that Duke employs gun thugs to intimidate them, and Kopple has plenty of footage of these armed men. Duke Power is not simply in the coal business, but in electricity and other power sources. The film says that 70 per cent of domestic coal resources are held by oil companies, so Duke is scarcely worried about a coal mine employing less than 200 people. But the confrontations with the police become more violent and the women decide to take action and turn up on the picket lines. They engage in direct action and lie in the street to prevent anyone going to work.

A subplot of the film also addresses events that interested popular media: the battle between Tony Boyle, who was President of the United Mine Workers of America (UMWA) from 1963 to 1972, and Joseph Yablonski, who challenged him for the presidency of the UMWA in 1969. The popular and dynamic Yablonski lost in what many people thought to be a rigged election. Later that year Yablonski and his family were murdered and Tony Boyle was arrested on a charge of having hired the killers. Boyle was finally convicted and sentenced to life imprisonment. These melodramatic events went on while the long strike continued. In the film Hazel Dickens sings the song 'Cold Blooded Murder' in reference to them.

At one of the pickets a young miner is shot dead by the gun-carrying management agents. This violent incident brings together the federal government, the union and Duke Power for discussions. After more than a year the strike ended with a new contract being accepted, although we are told that Duke Power caved in, not because of the dead man, but because the subsequent media interest in the story had reduced their share price. The strike was won, but soon most of the mines in the region would be closed as uneconomic. Today you can visit the Kentucky Coal Museum, as well as engage in a range of outdoor sports and activities. The 1930s tag of 'Bloody Harlan' has given way to tourist pursuits and sporting activities, but Kopple's film keeps the remarkable story of the strike alive.

One of the demonstrators at the picket lines in Harlan County carried a banner reading 'Remember Blair County'. This references the events that took place in the 1920s and led to one of the most important mining strikes in American history.

Matewan

Before we turn to the Battle of Blair Mountain we need to look at earlier events in the little mining town of Matewan. This was in 1920, when the United Mine Workers of America (UMWA) began to organize workers in the area and employees of the Baldwin-Felts detective agency arrived in the town and began to evict union miners from their homes. The mayor of the town, Cabell Testerman, and a sheriff, Sid Hatfield, with a number of men, tackled them and an armed fight broke out, at the end of which seven of the detectives were dead, as were the mayor and two of his crew. Perhaps unsurprisingly, the authorities found the coal companies free of any blame and put Sheriff Hatfield on trial for murder. He was promptly acquitted. A further trial followed this time on the charge of dynamiting a non-union mine, an activity that was not uncommon in these strikes. However, as Hatfield was walking into court he was shot dead by company assailants. This outrage provoked the formidable clashes that followed. The evictions continued and the displaced miners were forced to live in tents. Here, men firing machine guns and high-powered rifles attacked them. No detective, gunman or strikebreaker was ever prosecuted for any crime.

This is the story as it has come to be told, but what interests me is the fact that for years the incidents at Matewan, important though they were as a trigger for the dramatic events that were to follow, drifted out of public consciousness. The history of this struggle was not taught in local schools and the town sank into economic decline. Then, in 1987, John Sales released his feature film, *Matewan*, and the events of the past were recuperated. The town embraced its history in the form of a new kind of tourism. The National Park Service designated it a Historical Landmark, and it is now the home of the West Virginia Mine Wars Museum.

The film that brought this about begins with a narration that places the action firmly in the past 'back in them days' and is cleverly structured around the Western genre. It begins with a stranger coming into town and ends with a

FIGURE 36 Danny Radnor, Played by Will Oldham, in a tense moment from *Matewan*, 1987.

classic shoot-out. The incomer turns out to be a union organizer, Joe Kenehan, but, unlike the archetypal Western gunslinger, he is a dedicated pacifist, come to proselytize the virtues of collective action over violent exploits. In this he is out of sympathy with the traditional community values and the viewer is lead through an account of the town, its inhabitants and the dramatic events that follow by someone who is as new to all this as we are. Also on the train are a number of African Americans from Alabama, together with a group of Italians who have all been recruited to break the strike, even though they are unaware that their role is that of the scab. The train is attacked by a group of white miners who try unsuccessfully to stop them from reaching the town. Racial epithets abound and Sales manages in the early minutes of the film to establish the key themes that define the conflict. It explores unionism with its stress on camaraderie between people of different races and nations, but does not ignore the ingrained racism of the local community. It shows us that, although they are struggling to establish a union, patient collective action is not the preferred mode of a people who prize independent control of their own destiny and are sure that the armed and violent detectives who have followed Kenehan into town will be overcome only with countervailing aggression. One of the African American men is a union man and manages to persuade the others that he is to be trusted and so becomes instrumental in what follows. Sales mixes fictional characters with real historical figures: Sheriff Sid Hatfield and Mayor Testerman are here, and the events of the film do reflect the accounts of what occurred. Hatfield does prevent the eviction of a miner and is determinedly opposed to the lawless actions of the detectives. The strike takes place, the families move to a tented community, and they are attacked. Here, the Italian and American women who were formerly at odds come together, despite the lack of a common language, and the leading black character, 'Few Clothes' Johnson, played by James Earl Jones, is finally trusted and respected. However, he points out that in the event of violent action being undertaken, the black miners should opt out because blacks killing whites will be decidedly counterproductive, whatever the cause, and only whites take part in the final bloody battle.

Some of this is more invention than documented history, but several of Sales' imagined characters help in the structure of the film. A young boy, Danny, who is a preacher and keen union man, plays an interesting part in connecting daily life with the grand events that are unfolding, but also in using the discourses of the Bible and of religious services to inject into the industrial struggle the passion and language of the community. In a neat move, at the very end of the film, back in a coal mine where it began, the narrator is revealed to be Danny who is retailing the events he took part in.

The last major event of the film is the shoot-out between the Baldwin-Felts agents, the sheriff and his posse of miners. We know which of the historically accurate characters is going to die in this, but we transcend the code of the Western when the hero, Kenehan, is gunned down, so doesn't get to take the lonesome train out of town.

One of the virtues of the film is the way in which it makes us aware of the almost intractable position of the miners, caught between a remote union structure that is slow to send material help, and a constant threat from gunmen who act freely outside any kind of law. For the newcomers trying to make a living in a strange land this is particularly difficult. Defined as 'scabs' and therefore fair game to the miners, they are also bullied and threatened by the mine managers. As one of them says, 'we join the union, they shoot us. We don't join the union, you shoot us'. 'Well', says Kenehan, rather taken aback, 'That's one way of looking at it'. The cinematography was by Haskell Wexler, who filmed it in dark and muted colours: black and grey and green and blue. The soundtrack is composed of Appalachian music of the 1920s written and performed by Mason Daring. It is spare and plaintive and includes, to great emotive effect, the singing of the well-known West Virginia bluegrass singer Hazel Dickens.

The Battle of Blair Mountain

We have seen that coal mining and strikes are close companions, and when a mining industry emerged in West Virginia in the late nineteenth century, labour unrest grew up with it. There were strikes in 1892, 1894, 1895, 1897 and 1902. The last two involved violence, and at this time the coal owners enrolled operatives from detective agencies to act as strikebreakers.

As so often in mining areas, several modes of production existed at the same time and brought about different forms of subjectivity. An industry owned by large corporations was introduced into an area where the people had survived on small-scale farming, hunting, artisan mining and casual labour. A mining industry that for all its structured regimes of production, and need for a disciplined labour force, blatantly cheated its workers, housed them badly, paid them partly in scrip that was only redeemable in company stores and relentlessly drove down wages. Without the need to pay compensation for injury or death, and with a large surplus of available labour, miners were, as so many said, 'cheaper than mules', and could be treated with contempt. Nevertheless, many independent-minded Appalachian mountain men resisted the suggestion that they should join a trade union, and when they did embrace the idea, and the UMWA began to organize workers in the area, a most famous conflict occurred. It became known as the Battle of Blair Mountain.

Perhaps the most succinct account of the struggle may be found on the West Virginia Historic Highway marker. It was erected in 2002 and was made by the West Virginia Division of Culture and History. It reads:

In Aug of 1921, 7000 striking miners led by Bill Blizzard met at Marmet for a march on Logan to organize the southern coalfields for the UMWA. Reaching Blair Mt. on Aug.31, they were repelled by deputies and mine guards, under Sheriff Don Chafin, waiting in fortified positions. The five-day battle ended with the arrival of U.S. Army and Air Corps. UMWA organizing efforts in southern WV were halted until 1933.

FIGURE 37 Machine gunners at the Battle of Blair Mountain, 1921.

The union workers were armed and their intention was to move through the un-unionized counties of West Virginia organizing the workers and repelling the company gunmen. This provoked the clashes that grew into the largest civil insurrection in the United States since the Civil War. The miners were Appalachian hill folk, African Americans who had migrated from the south and European immigrants especially from Poland, Wales and Italy.

At this point a central character becomes important, that of the sheriff of Logan County, Don Chafin, who recruited local men to try to halt the march. He was very successful in this, as people flocked to try to end the rebellion. At the same time, both sides were actively recruiting. A respondent to a government survey in 1934 said that:

> The operators have these gunmen to do their shooting for them. One time when they were fighting at Blair Mountain I was walking along the road and a man comes up to me and says, 'Say, do you want to make $10.00 a day?' And I says, 'Ten dollars a day doin' what? "Fighting at Blair Mountain; you'll get $10.00 a day and all your expenses". They were looking for men everywhere.' (Morris 1934: 141)

Around 3,000 men finally joined in the Battle of Blair Mountain. It lasted for less than a week and as a battle was rather a low-key affair. Despite the unique event of the miners being bombed from the air, there was little close combat. It was hard to see through the dense undergrowth, and many combatants never got a good look at the opposition. It ended when federal troops arrived and the miners, many of whom had fought in the First World War, laid down their arms rather than fire on soldiers with whom they had no dispute. Although the best estimate is that sixteen people died, a dozen of them miners, the importance of the battle is less the drama of the fighting than the effect it had on the workers' movement generally.

In the short run it had deleterious consequences for the mining unions. They lost many members, and public sympathy lay with the mine owners. But over time it became a key memory in American public life and an event that helped to question the belief system of the country and its sense of being an equal and democratic place, free from the internal divisions and conflicts that were common in Europe. For, while race was inevitably considered to be a divisive issue, overt class warfare sat uneasily within the American ideology of growing prosperity for all through the advance of capitalism. The battle was frequently represented in newspapers and magazines as the typical exploits of hot-headed mountain men who quickly resorted to violence and were deaf to rational argument. This allowed the problem of class to be given a spatial dimension and shuffled off to socially and politically remote places in Appalachia. At the same time, within the American imaginary, the mountain people were viewed as tough, pioneering types who could take care of themselves in a fight, and exemplify the national virtues of sturdy independence and self-reliance. Although this was clearly a class struggle, the workers were mobilized by many grievances, but still subscribed to the dominant ideology of the country. Robert Shogan has commented on this:

> The Blair Mountain uprising demonstrates that, middle-class mythology to the contrary, class conflict does exist in America. But it also illustrates the limitation of that conflict in this country. Standing between the miners of West Virginia and an outright working-class revolution against the Federal Government was their hope for their country based on its promise of opportunity, individual freedom and fairness under law. (Shogan 2004: 222)

Looking at photographs of the struggle one is struck by how unlike a military engagement it was. Images in the West Virginia State Archives show men in working clothes and overalls with serious faces and antique rifles. There is nothing to identify them as miners as they line up at the end of the engagement to hand in their guns. There are also photographs of the federal troops looking bored and lounging on the ground waiting for the whole thing to be over.

Compared to the long struggle for union rights and better conditions in mining areas, these skirmishes were relatively unimportant. But the ways in which they have been represented bring mining into the mainstream of American culture. They stress rugged independence, and an armed attack on the forces of reaction.

FIGURE 38 Miners handing in guns at the Battle of Blair Mountain, 1921.

They call up the pioneer spirit and locate mining as one of the principal fields on which the battle for social justice has been fought.

US trade unions still find it difficult to recruit in mines especially as new forms of mining – strip mining and mountain top removal – have changed the nature of the work and weakened the sense of collectivity so central to the culture of mining. In the next chapter I look at these technologies and at some of the ways in which they have been pictured.

6 THE NEW LANDSCAPES OF COAL

Coal throws up many curious and intriguing stories, not the least of which is that of the famous underground fire at Centralia, Pennsylvania. This was a mining town that by the 1960s was a flourishing place with churches and banks and good homes and schools. Some 2,000 people lived there until in 1962 a fire broke out on a coal tip, and this set ablaze the anthracite below ground. It is estimated that there is more than 20 million tons of this coal and it continues to burn steadily, despite all the efforts of the fire brigade and the US Bureau of Mines to extinguish it. It burnt for decades, getting steadily hotter and smokier at the surface, making life increasingly difficult. In the 1980s the federal government paid to evacuate the town so that today only a handful of people remain, and the town itself has been bulldozed out of existence. The popular writer Bill Bryson provides an interesting description of Centralia in his book *A Walk in the Woods,* an account of his time walking part of the Appalachian Trail (Bryson 1997). A more nuanced and politically charged version of the story is given in Joe Sapienza's 2017 film, *Centralia: Pennsylvania's Lost Town.* Because of its burning slag heap the town has become very famous and is one of the most visited tourist sites in Pennsylvania.

Burning coal tips are one way in which coal deforms the landscape. Another famous one at Jharia, India, has burnt for more than a century, but most slag heap fires are small and containable. A permanent, steady blaze like the one in Centralia is clearly rare, and its very novelty is what draws the tourists, photographers and photojournalists to it. Fires of this kind, though, also remind us that the effects of coal mining may continue long after the last ton of coal has been mined. Around the world many billions of pounds have been spent trying to clear up after mining; while, in other areas, the dereliction remains. This is not simply a matter of making good the devastation of previous centuries, for mining is still a booming industry and, because of technological advances, now causes more damage than was dreamed of in the past.

One example of this is given in Zhao Liang's film *Behemoth* which provides a poetically charged look at mining and iron working in Inner Mongolia. The stunning natural beauty of the place as sheep graze in emerald green fields is being slowly destroyed by the march of an opencast coal mine. The film is structured in the form of a terrible dream, and draws on literary and mythic sources. It also has considerable documentary value. It patiently records the appalling labour of the miners and dwells on the scars and mutilations of their bodies, which mirror the defacement of the landscape. At the end of the film Zhao Liang takes us on a journey to see what is being made as a result of all this painful toil and destruction of the natural landscape. Lines of high-rise buildings, many of them entirely unoccupied, fill the scene. These 'ghost towns' were built for large populations who have yet to show up and are the urban future that will overshadow the scarred Mongolian land.

In South Africa there are more than 6,000 abandoned mineral mines, spilling acid water and heavy metals into the environment. The major mining companies are said by Actionaid to have little regard for water quality, and they pollute rivers and streams with impunity. In addition to the contamination of groundwater from abandoned and sinking mines, greenhouse gases and air pollution are grave problems, as are mining accidents that lead to serious injury (Actionaid 2014). In China, too, some industries are being run down, and the death throes of once prosperous industry are as protracted as those in Western countries. Wang Bing's extraordinary documentary *West of the Tracks* lasts for nine hours and slowly explores rustbelt China. It is set in a district of Shenyang, which for half a century was China's largest industrial base, and thirty years ago more than a million people were employed there. By the start of the new millennium it had shared the fate of so many centres of heavy industry in the West and was falling into economic and physical ruin. Wang Bing shot the film without a crew on a small digital camera. It explores the rust and ruin of the industrial works as well as the decay in the lives and settled way of existence of the workers and their families. The critic Lu Xinyu writes of *West of the Tracks* that it is 'without question the greatest work to have come out of the Chinese documentary movement, and must be ranked among the most extraordinary achievements of world cinema in the new century' (Xinyu 2005: 128).

This is very high praise, but the film – documenting violent social change, the collapse of industry and the dissolution of the collective spirit that it engendered – is certainly remarkable. As we move through the ruins in a long slow take, it is worth remembering that the world is full of marginal areas where normal life is blighted by huge zones of wasteland.

One of the most famous is the 'Black Triangle', a term used to describe the borderland between the Czech Republic, Germany and Poland. It earned the title in the 1980s because of the extraordinary amount of pollution it generated, largely from the mining of brown coal. In 1994 the distinguished photographer Josef Koudelka produced a photographic report on the Ore Mountains in the region. He used a panoramic camera and the long, thin prints do not allow us to take in the landscape at a single glance. Rather we have to scan across the pictorial frame and study it section by section. What we see is a landscape of remarkable desolation. The black-and-white images are empty of people but full of the actions of people who have created waste heaps, craters and long straight, ugly concrete roads that cut through screes of coal and dust to vanish on the horizon. This book of photographs was influential in drawing attention to the region, and to the whole subject of post-industrial landscapes (Koudelka 1994).

Sometimes the landscape that is being degraded has particular spiritual or cultural meaning to a social group. The Australian experience is explored in João Dujon Pereira's film *Black Hole: Transforming a Forest into a Coalmine* that was premiered in 2015. It tells the story of a fight to prevent the expansion of Whitehaven Coal's Maules Creek mine into the Leard State Forest in New South Wales. This is a complex story, at the heart of which is the destruction of a site of outstanding natural beauty in order to develop a strip mine. One source of opposition to the plan is the Aboriginal people who regard it as a sacred site. They join forces with local farmers and environmental activists to oppose the project. They all receive training in non-violent direct action, and learn the need to develop patience as police and local officials routinely harass them. The activists charge the coal company with corrupt practices and local and national government with aiding them. Certainly, the major coal owners in Australia are anxious to avoid clashes of this kind and, we are told, go to considerable trouble to make mining seem like just one more natural activity in rural settings. The aim is to make it seem like an activity that can easily coexist with farming and, as the mines encroach on Hunter Valley, with wine making.

Even in places where mining is carefully regulated and subject to controls on its activities, the surrounding areas inevitably bear the social, economic and environmental impact of the industry. So, many groups and organizations have looked at ways of restoring the blighted land to a state of nature. Collieries are cleared away, parks and nature reserves are created, slag heaps are seeded, trees are planted and nature is restored. The idea that wild places can be constructed on the scarred land of the mine, however, has often been criticized. Commenting on this notion of 'rewilding', Ted Nield says:

> Today, defunct opencast mine workings are sometimes replaced with artificial re-creations of 'natural' geomorphologies deemed right for the location by expert landscape designers and archeologists. I suppose one could call this 'geomorphological rewilding', creating a 'stage-set' landscape and imposing a new dimension of artificiality on what is already artificial. The concept of wildness, in Britain at least, is little more than romantic nonsense. (Nield 2014: 71)

Post-industrial landscapes and photography

By the time of the miners' strike in 1984 the mines of south Wales were becoming increasingly uneconomic. As each one closed, workers were made redundant or had to move to other pits further away from their homes. Sometimes miners moved several times and it was clear that the industry was slowly dying. The strikers wanted a slow and rational process of closure accompanied by the development of new industries and economic activity. It was against this background that one of the most significant photographic surveys of a British post-industrial society took place – *The Valleys Project*.

From the last decades of the nineteenth century there were many photographic surveys of areas in the British Isles. Although amateur photographers carried these out, a range of organizations such as libraries and museums supported them. In the 1880s and 1890s in Wales, surveys were imbricated within other movements that sought to define the particular culture of Wales as separate from the rest of Britain. The University of Wales received its Charter in 1893; the National Eisteddfod was revived and was supported by the Honourable Society of Cymmrodorion, which promoted Welsh arts and science. At the same time, there was a movement to establish a National Library and a National Museum. Vena Louise Pollock, who has researched all this, tells us that the photographs produced by the survey of Glamorgan tended to stress the ancient, and paid scant attention to the industrial life of the county that was booming at the time (Pollock 2009). A particular conjunction of national aspiration, the founding of powerful organizations and relationships with people carrying out surveys elsewhere in the world all influenced the nature of the work that was produced.

The Valleys Project

Moving forward a hundred years to 1983, *The Valleys Project* was initiated. It was a photographic survey commissioned by Ffotogallery, which was founded in 1978 and was the first organization dedicated to systematically showing photography in Wales. In addition to exhibiting photographs, it commissioned work, supported artists, toured shows and produced exhibition catalogues, a magazine and books. In 1983 its then director initiated *The Valleys Project,* and a year later its magazine *ffotoview* announced the fact that an exhibition of the first commissions for this project would be held at the gallery. The photographers were Ron McCormick and Paul Reas, while the local council supported the work of John Davies. They were followed in subsequent commissions by Mike Berry, Francesca Odell and Roger Tiley. A detailed account of the project is given by Paul Cabuts in a book on Welsh photography (Cabuts 2012). He stresses the importance of Newport School of Art and Design in getting the project off the ground and contextualizing it both within the contemporary scene and the history of photography in Wales. The founding at Newport of a documentary photography course, with the Magnum photographer David Hurn at its head, gave impetus and focus to documentary photography in Wales. From the end of the 1970s the School began to undertake photographic investigations – *The Newport Surveys* – which were thematic explorations of life in south Wales. These became very well known and were clearly an influence on the *Valleys Project*, especially as former students of the School were among the commissioned photographers.

FIGURE 39 Mike Berry, *Hauling Coal Back to the Village, 1985.* Scrabbling for coal in the miners' strike. From The Valleys Archive.

The Valleys Project was undertaken within a particular conjunction of influences. The direct impact of Newport School of Art was matched by other factors. It is worth noting that the first phase of the project took place while the miners' strike was in full swing, a fact that was central to the photographer's relationship with the valleys, and a key to understanding the nature of the images that were produced. Interestingly, the photographers were exploring the nature of what de-industrialized valleys might look like, and how they might function, while coal mining was still in existence. At the same time the heritage industry was growing. Big Pit Blaenavon had opened as a tourist attraction in 1983, while just a mile or so away there was a functioning coal mine. This Janus-headed approach is quite common within coal cultures, but it does problematize the way in which we can describe the present or anticipate the future.

The intellectual climate of the time was also important. A new school of social and cultural historians were re-assessing the history and importance of industrial south Wales. Both Dai Smith and Hywel Francis, who were key figures in this movement, wrote introductions and commentaries to the publications of the Valleys Project. Roger Tiley's documentary photographs directly addressed the miners' strike with images of strikers, support groups and pit villages. In the catalogue that accompanied the exhibition, his work is framed by Hywel Francis' passionate article. Mike Berry worked in Glyncorrwg and presented a picture of a community still full of warmth and vitality despite the social and political problems of the time. Similarly, Francesca Odell looked at Clydach Vale, especially at the lives of young people in the village.

FIGURE 40 Francesca Odell, *Street Corner, Clydach Vale, Tonypandy, July 1985.* From The Valleys Archive.

Paul Reas concentrated on the new technology firms that were moving into the valleys as part of the attempt to look to the future and provide employment other than in coal mining. Among the documentary work the project also commissioned Peter Fraser who worked in colour, showed his images in galleries and was outside the documentary tradition. Describing his work in an interview with Sue Beardmore, he said:

> What convinces me that photography is a fine art is that it has demonstrably the same capacity as other fine arts such as painting and music to express the full range of one's spirit and that is precisely what I am trying to say – we have hardly begun to explore the expressive potential of the medium, but in this country in particular, we are still following very set patterns in terms of imagery. We try to make explorations in spite of the all pervasive, documentary, utilitarian influences, but progress is slow. (Beardmore 1986: 21)

The presence of Fraser's work within the archive broke the hegemonic use of documentary photography and hinted at other, more allusive ways, of picturing mining communities. But another of the contributors to the project was the well-known fashion and portrait photographer David Bailey. He volunteered to take part having read about the project and was excited by the idea of the endeavour, especially its relationship to other major archives such as the FSA. Bailey was a stranger to the area, as were some of the other photographers, and he concentrated on representations of the built environment. He printed his images in a very dark and sombre monochrome, and showed terraced housing, collieries, smokestacks, churches and shop fronts. Unlike the other photographers, who had worked closely with the local communities, Bailey stood back from the social life except as it was expressed in the physical surroundings.

Reconnaissance – Wales

Czechoslovakian-born Josef Koudelka trained as an engineer, but gave up this profession to become a full-time photographer in 1967. At that time he was photographing gypsies, a project that engaged him for many years. The Russian invasion of Czechoslovakia in 1968 became a defining moment for him as he recorded the assault of the troops in Prague, smuggled his negatives out of the country and had them published in the *Sunday Times*. They became the definitive photographs of the invasion and received a number of awards. In 1970 Koudelka was given political asylum in Britain, where he lived for a number of years while travelling extensively in Europe.

He has made a study of many desolate landscapes in Greece, France, Lebanon as well as new work in what was then Czechoslovakia. In large panoramic photographs he recorded places that had been degraded by war and the toxic traces of industry.

In the late 1990s *Ffotogallery* invited Koudelka to visit South Wales, where, over a period of two years, he photographed industrial sites, opencast mines and land reclamation sites. The result was *Reconnaissance*, a collection of remarkable images that was published in 1998 (Koudelka 1998). The large panoramic pictures are extraordinarily detailed and allow the viewer to read the often-bleak landscapes without being directed to a particular point of view. At the same time they are beautifully balanced and full of a kind of elegant design. Empty of people, the land is absolutely marked, sometimes scarred, by human activity. He also photographed Tower Colliery, the last working mine in South Wales, but this image is the most factual and pedestrian. It shows a pit head and the jumble of artefacts that crowd the surface of a mine. For the most part he produced an abstracted landscape elegantly presented. Critics of the work have seen the carefully composed images as telling us little about the social conditions and economic prospects of the place. Those who admire it praise him for drawing attention to otherwise abandoned places and working outside the conventions of picturing mining country, seeking neither the picturesque nor obvious markers of industrial dereliction. It is instantly identifiable as Koudelka's work, for he has found and recorded places of this kind around the world. It is a body of work that is very different from the *Valleys Project* with its predominantly documentary impulse and desire to map not only the topographic, but also the social and cultural features of what was becoming post-industrial Wales.

Stripping the land

While post-industrial landscapes are being neglected, cleaned up or transformed into sites of leisure, mining is still creating dereliction at a very rapid pace. One approach to the story of mining is to follow its history of constant technical developments. Machines have been created that speed up processes, boost productivity, cut the numbers of workers needed to get the coal out or transform what we have traditionally understood by 'mining' altogether. While underground mining has been subject to such changes throughout its history, surface mining (strip mining) dramatically increases the use of machinery and radically changes the nature of what we understand by 'mining'. It alters too the relationship between mining and the natural landscape. Strip mining takes place in many countries and consists of ripping away the surface land, clearing away trees, vegetation and many feet of topsoil to reveal seams of coal that lie close to the surface. Although coal has been extracted from near the surface for centuries, the serious practice of mechanized strip mining has been developed since early experiments in the 1920s and now uses some of the largest machines on the planet. Of course, this kind of coal gathering employs many fewer people than deep mines and the workers are not miners in the traditional sense. For the most part they manipulate machines, and are frequently mocked by old-style colliers as being mere sedentary drivers.

FIGURE 41 Garzweiler Strip Mine, Germany, Bildagentur Zoonar GmbH. Original in colour.

Once all the coal has been scooped out of a site the land has to be restored in one form or another. However, many ecologists argue that no real regeneration is possible given the dramatic effect the practice has on the flora, fauna, watercourses and the land it has torn open. It is no surprise to learn that wherever strip mining takes place you will find an active environmental group that opposes the practice. Sometimes these include people living in the area who are appalled by the streams of trucks coming and going to the site, as well as by the constant haze of dust that lies over their houses, and the despoliation of streams and rivers. But objections are also raised by large and sophisticated organizations that deplore the long-term effects of this kind of mining on water supplies, the local habitat and environment, as well as the health of people living in the surrounding communities.

While ecological groups draw attention to all these drawbacks, many countries carry on with the practice. About 80 per cent of Australian coal is mined this way, and the United States has numerous large strip mines. Most British opencast mining takes place in Scotland, a country with a long mining history, although the industry is declining there. Britain's largest opencast mine is on land to the north of Merthyr Tydfil in a traditional mining area of South Wales. The grandly named 'Ffos-y-fran Reclamation Scheme' aims to remove 11 million tonnes of coal in seventeen years, before converting the land to residential use. Unfortunately the land is already in residential use as the site is very close to local houses. Residents make the usual complaints of disruption by noise and of air pollution, and they also report a high incidence of childhood asthma and of cancer clusters. They also point out the proximity of schools and playgrounds to the site, which is the size of 400 football pitches. The mine's operator, Miller Argent, argue that the land was already degraded and pocked with old mine works, and that they will responsibly restore it to heathland that can be built on. In 2017 the United Nations called for an independent enquiry into the potential health impacts of the site. Other European countries also permit this kind of mining to exist. Despite Germany's declared intention to close down both coal and nuclear plants in favour of renewable resources, the biggest mining operation on earth is sited there. The Garzweiler strip mine in North Rhine, Westphalia, covers some 18.5 square miles and aims to extract 1.3 billion tons of dirty, brown lignite coal during the years of its operation.

Mountain top removal

Mountain top removal (MTR) is often called 'strip mining on speed'. It is very much an Appalachian practice and began there in the 1970s. In this sacrifice zone more than 400 mountains have been devastated. As with strip mining, MTR involves clearing the trees and plants from the surface of the earth, then removing up to 500 feet of soil to get to the thin seam of coal beneath. This top layer is called the 'overburden' or 'spoil', and it is blasted away by explosives so that the entire top of the mountain is removed. Most accounts say that the spoil is often simply dumped in the valleys (or hollows) below, although the managers of these schemes claim

FIGURE 42 Mountain Top Removal at Oven Fork, Kentucky. The photographer wishes to remain anonymous. The original is in colour.

that they deal with it responsibly. The coal tip is here inverted, so that, instead of a mound of debris, the spoil falls away to choke streams and valleys. Opponents of MTR calculate that some 500 mountains, among the oldest on earth, have already disappeared; more than 3,000 streams have been choked with filth, and whole communities have been displaced, their people forced to move as the land is bought up by private companies, and the hill blasted clean away.

In order to be sold, the coal has to be cleaned, and this process leads to the creation of huge dams of toxic sludge containing mercury, lead, arsenic and chromium. There is always the possibility of accident. In 1972 an impoundment dam in West Virginia burst and 130 million gallons of polluted sludge swept into the creek below. More than a hundred people were killed outright with a thousand injured and 4,000 rendered homeless. But even where there are no dramatic incidents, the slurry soaks into the ground and may contaminate water and air supplies. Several reports have noted an increased rate of heart attacks, cancer and lung disease in nearby residents, but supporters of mining have argued that this increase is a consequence of the lifestyle of the inhabitants rather than external environmental factors. As so often in mining country, the profits from all this devastation flow out of the region, leaving Virginia, West Virginia, Tennessee and Kentucky as among the poorest states in the Union. The natural beauty and biodiversity of Appalachia are destroyed wherever MTR takes place, and it is worth noting that this kind of mining employs very few people and brings little community benefit.

MTR carries great symbolic resonance in Appalachia, a region that occupies an enigmatic place in US culture. Its residents have few life choices: as a character in Michael Apted's 1980 film *Coal Miner's Daughter* (a biopic about Loretta Lynn) puts it, your choices are 'coal mine, moonshine, or move it on down the line'. Appalachia is, of course, the home of mountain folk, respected figures for their toughness and independence. It's known for its mines and miners, down-to-earth hard workers with a reputation for strength and camaraderie and who are the backbone of tight-knit, family-based communities. Famed for its music, it is also stigmatized as the home of feckless hillbillies, uneducated, unemployed and getting by on public handouts. These contradictory positions are in constant tension within the culture of a place that has remained economically depressed for generations, as mining died out and was not replaced by other industries. The American dream of steady progress and increasing wealth has not been manifested in large parts of the vast region. Rebecca R. Scott notes the whiteness of Appalachia, but also that the people have not prospered in line with the notion that whiteness is expected to ensure steady economic and social advancement. Geographically and socially marginal, and existing outside the mainstream of American life, the hillbillies, she observes wryly are 'white people who fail to be white enough' (Scott 2010: 109).

MTR seems to endorse the most negative ideas about the region. Over a huge area, much of it of great natural beauty and biodiversity, destruction and environmental contamination are allowed to occur. Protestors argue that this is because it is seen as a cheap, degraded place that lacks political clout. What power it did possess in the mining industry disappeared when the coal companies fought and broke the unions, leaving them with a free hand to, quite literally, carve the place up. An important text in relation to mining in Appalachia is that of Ken Light and Melanie Light. In the introduction to the book, Melanie Light notes that:

> The mountain folk of Appalachia are a cultural paradox for much of America. On the one hand they have been mythologized in a nostalgic way and acknowledged as the source of much American folklore and culture. On the other, they are seen as shameful 'trailer trash', ignorant hillbillies. But, in truth, most Americans do not think about Appalachia at all, sensing that it is caught in an unfashionable past, which is worse than being dead. (Light and Light 2006: 1)

Oral histories are an important source of information about mining communities, as historians tried to find out something of the ordinary life of these places which otherwise remained mysterious. The accounts in *Coal Hollow* allow us to hear a range of voices – a journalist, a retired coal miner, a woman described as 'a mountain woman', a preacher, a mayor and so on. Reading them one is struck by the constant change that these apparently fixed and settled communities undergo. Writing of his early life, Ernest, aged 46, says:

After the last bust, people left and right were moving. Dad couldn't move because he was disabled. You'd see people loading their cars, and they'd take off. You'd see 'em, just leaving. They'd leave their houses with their furniture in it. Most people from this area went to Chicago. All kind of kids I grew up with – they'd be here one day and the next day they was gone. (Light and Light 2006: 127)

Ken Light's black-and-white photographs capture the atmosphere of the stable community, but there is a febrile quality to many of the images that injects them with a restless feeling. Folk sit on their porches; children play in the dirt and swim in creeks. There are numerous portraits of people with lined and characterful faces, and shots of religious ecstasy in revival tents. There are dark images too: a man cradles a fighting cock, Klan members listen to speeches, a simple headstone is dedicated to *Baby Girl Strunk, Born and Died August 1951*.

Despite its long history of poverty and disadvantage, Appalachia still resonates in American popular culture. For example, John Grisham's 2014 legal thriller, *Gray Mountain*, is set there and tells us a good deal about everyday existence in coal country. Strip mining has devastated the land and left streams and rivers contaminated, so it is not advisable to drink the water. Lakes of polluted slurry leach into the ground and threaten to flow into the water supply. Over twenty years a thousand miles of headwater have disappeared. Near the strip mines the ground shakes, cracking cellars, while above ground coal dust fills the air. The people still living there are subject to high rates of cancer, while miners and former miners succumb to black lung (pneumoconiosis). Crystal meth is the drug of choice, together, of course, with alcohol. The mining companies, who are reckoned to have caused this devastation, are unscrupulous in ways familiar to miners for centuries. They vigorously fight compensation claims for industrially caused ill health. They pursue the radical reformers who are the main protagonists of Grisham's novel and bring in the FBI to limit the scope of their actions. Because they have driven out the unions, they are able to sack anyone they dislike without having to give reasons for their actions, and regularly find ways to underpay the workers. In short, Big Coal still runs the place. It is a picture that would make perfect sense to Upton Sinclair who covered some of this ground in *King Coal*. Even the fact that the companies are located outside the region would be familiar to him, although he would not expect members of the Russian Mafia to be in their ranks. Meanwhile, the towns are dying, their buildings stand empty and the once-beautiful landscape has been destroyed by strip mining and mountain top removal (Grisham 2014).

In 2016 a memoir on Appalachian life became a perhaps surprising bestseller in Britain as well as in the United States. J. D. Vance was brought up in the once-prosperous, now-decaying steel town of Middletown, Ohio, but his family came from Kentucky. In the book he looks back at his troubled family history and the complex web of violence, alcoholism and drug addiction that marked his childhood. But this is no cry of deprivation for he also lauds his family and the community within which they lived for their generosity of spirit, loyalty and intense family feeling. Education is the route by which Vance got out of hillbilly country. He worked hard and attended Ohio State University and Yale Law School. A spell in the Marines followed, and he ended up as a Republican voting Silicon Valley investment manager. In other words, he became a fine example of how to live the American dream. The heart of the book, however, is the community from which he sprang and when it topped the *New York Times* bestseller list it seemed to be a key to understanding the white underclass that has been generally demonized or treated with contempt by the dominant American society. These were the mountain people, the coal miners and the terminally unemployed who were courted by Donald Trump in the run-up to the 2016 election. With the election looming, the well-educated Vance interpreted the lives of the downtrodden and dispossessed to cable TV audiences. He doesn't defend this culture; indeed, he finds it toxic and thinks that the community needs to lose many of their deeply engrained cultural norms. They also, he contends, need to break the habits of drug taking, eating junk food, groaning against injustice and being shiftless. He frequently hands out good advice to his erstwhile community:

> We choose not to work when we should be looking for jobs. Sometimes we'll get a job, but it won't last. We'll get fired for tardiness, or for stealing merchandise and selling it on eBay ... We talk about the value of hard work but tell ourselves that the reason we're not working is some perceived unfairness: Obama shut down the coal mines, or all the jobs went to the Chinese. These are the lies we tell ourselves to solve the cognitive dissonance – the broken connection between the world we see and the values we preach. (Vance 2016: 147)

The book came out at a particular moment, just before a presidential election, and it raised all kinds of questions about the nature of ingrained poverty and the role of the state in making social change. The election of Donald Trump, not least because of 'hillbilly' votes, meant that politicians and political pundits took it very seriously. It marks a shift away from regarding ex-mining and heavy industry communities as unfairly depressed, to blaming the poor for their current state. Vance is more interested in personal conduct and group values than he is in the structural economy. It is very much a white underclass that he describes, but their problems may be found in many other regions around the world.

Mountains and the sublime

MTR puts industry in direct conflict with natural life, and mountains, once regarded as dark, dangerous, uninhabited places, have for centuries been seen as sites of spiritual seclusion, contemplation and natural beauty. In Chapter 1 I noted

Edmund Burke's observations on the nature of the beautiful and the sublime. While the beautiful appeals directly to our aesthetic sense, the sublime is associated with pain, danger and a particular kind of delicious terror. These powerful emotions provoke in us feelings of awe and astonishment as well as deep reverence and respect. One trigger of the sublime is vastness and Burke comments:

> Extension is either in length, height, or depth. Of these the length strikes least; a hundred yards of even ground will never work such an effect as a tower an hundred yards high, or a rock or mountain of that altitude. I am apt to imagine likewise, that height is less grand than depth; and that we are more struck at looking down from a precipice, than at looking up at an object of equal height; but of that I am not very positive. (Burke 1757: 57)

Photographs of MTR are invariably taken from above looking down at the broken land and, in keeping with Burke's suggestion, inspire particular feelings of emotional vertigo. Burke's essay was critiqued and refined by Kant, but has been extraordinarily influential in determining how we should respond to particular landscapes. It is clear that MTR is far from beautiful, but it is certainly sublime, evoking in us feelings of amazement and terror. It is, in Burke's sense, truly awesome.

The sublime in art

In the eighteenth- and nineteenth-century landscape, painting was a particularly important genre in both Europe and America. The forms of painting based on principles articulated by artists and philosophers – the pastoral, the picturesque and the sublime – all had their adherents. Pastoral images showed a peaceful landscape dominated by human beings who had tamed nature to produce elegant gardens, verdant agricultural fields and tranquil walks and lanes. The sublime was, of course, associated with nature 'red in tooth and claw', nature that overwhelmed the senses and filled the viewer with a delightful dread. Seascapes showed towering waves and storm-tossed boats; mountain ranges were magnificently vertiginous; avalanches and volcanic eruptions were ideal subjects of study. Joseph Wright of Derby made several pictures of an erupting Vesuvius; Philippe-Jacques de Loutherbourg depicted war at sea and, more pertinently to our study, works of the industrial sublime of which *Coalbrookdale by Night* (1801) is perhaps the best known. J. M. W. Turner studied sublime subjects with storms at sea, but also experimented with the depiction of rain, fog, water and fire.

In Germany the most significant figure was the Romantic painter Casper David Friedrich. In the United States the nineteenth century produced many celebrated landscape painters. At first they followed within the tradition of the British school, but the particularity, the size and the grandeur of many American landscapes began to lead to the production of a particular set of concerns and aesthetic approaches. The English-born Thomas Cole was a founder of the movement, but Americans such as Jasper Francis Cropsey, Frederic Edwin Church, Albert Bierstadt and Thomas Moran established a new way of looking at these landscapes.

The American technological sublime

The tradition of painting the sublime was, then, well established in the United States by the end of the nineteenth century, but this period also marks the rapid development of American commerce, trade and, most significantly, industry.

In an extremely influential book, David E. Nye argued that the source of feelings of sublimity, once evoked by the depiction of nature, moved to awe generated by feats of engineering and technology. Where once the sublime was to be found in works such as Thomas Moran's view of the Grand Canyon, or Frederic Edwin Church's tempestuous scene of Niagara Falls, later Americans were moved to these emotions by great feats of technology. The sublime was evoked by mighty bridges, by the new and astonishing skyscrapers, or images of terror – the atomic bomb. Following the work of Perry Miller and Leo Marx, Nye traces ideas about the sublime through a study of technology. The book is organized chronologically so that, in a way, it offers us a history of the sublime as it moves away from nature to the technical achievements of the nineteenth and twentieth centuries. Each spectacular technological feat drew crowds of Americans to view them and, more importantly, became the objects and events that articulated a new,

FIGURE 43 A stock photograph showing the power of the Hoover Dam.

decidedly American, culture. The technological sublime, then, was important in the collective psyche of the United States. The power of machines and the clean aesthetic of modern technology represented the very embodiment of progress and enlightened development. They also changed the way in which industry might be depicted. Sharon Zukin has praised the way in which industry was represented by leading artists:

> As early as 1908 the Pittsburg Survey, financed by the Russell Sage Foundation, commissioned artist Joseph Stella to document conditions in the city's steel mills. Instead of the loss of human dignity that had been shown by photographer Lewis Hine in coal towns and textile mills, Stella showed the power and beauty of steel's large-scale facilities. This set a model for artistic representation of twentieth century economic power. Fifteen years later the photographer Edward Weston – now known for his detailed studies of natural forms and nudes – documented the buildings of the Armco steel plant in Ohio … From the late 1920s until 1936, Margaret Bourke-White made dramatic formal studies of industrial sites. (Zukin 1991: 63)

The Pittsburgh Survey was an important sociological study of working class life in one town. In fact, Hine was the staff photographer on the survey that combined academic investigation with political critiques. The iconography of industrial drawings and photography was very complex. Stella's drawings are vibrant and display the power of industry, but, at the same time, there was still the secret world of dirty, greasy labour, not least in the mines. Despite the aesthetic gloss given to industrial processes by artists and photographers, Bob Johnson points out that:

> As a cultural object, coal served as a reminder to industrial-age Americans that beneath the surface of the nation's veneer of steel, glass, and light and beneath the exhilarating speed at which its factories, ships, and railroads churned, there always stood this other implacable, bare-faced, and hidden material world with its furnace stokers buried in the depths of steamships, its miners sequestered in mountain hollows, its dirty ashes hurried out the backdoor, and its smoke and slag removed beyond the sight and smell of the nation's middle classes. Nothing was darker, dirtier, or more decidedly modern than coal, and nothing evoked so explicitly the traumas of modernity as did this gritty black rock and the trace of its lingering smoke. (Johnson 2010: 266)

Johnson reminds us that at the bedrock of the glittering technological sublime lay seams of coal in places hidden from sight. But, today, increasingly, the deleterious effects of coal and other carbon-based fuels are being clearly understood; so much, indeed, that some people argue that we have entered a new epoch, one that needs to be named in a way that acknowledges the impact of human beings on the entire world.

The Anthropocene

In some ways the technological sublime encouraged us to turn our gaze away from nature, which was often seen as coming fresh from the hand of God, or representing powerful spiritual impulses, to an inward regard of our own achievements. This may be in keeping with the times in which we now exist. For, while according to geologists we are living in the Holocene epoch, which dates from the last ice age, some 12,000 years ago, there is a growing movement among scholars and environmentalists to label our current epoch the Anthropocene (from the Greek *anthropos* for 'humans' and *kainos* for 'new'). Human beings, the argument runs, have had such a profound effect on the natural world, polluting oceans, killing off entire species, reshaping the landscape, building powerful dams, testing nuclear bombs and so on. These acts have been so deep and serious that we may consider them to have to brought about a new era. The word 'Anthropocene' has become extremely popular in environmental circles because it stresses the activity of humans in changing the nature of the natural world. Many geologists disagree, arguing that a new epoch would require a dramatic shift, one that can be read at the structural level of the rock strata. They see the title 'Anthropocene' as little more than a political slogan that systematizes the diverse ways in which we interact with the planet in a harmful manner. Also, others argue, it is too broad a concept for any kind of serious analysis of the problem, as it does not distinguish between the people and institutions that drive global pollution, from those that are merely victims of it. The word was introduced to the language by the chemist and Nobel Prize winner Paul Crutzen at a conference in the year 2000 when he called for it to be used to define a new human-dominated geological epoch. He dated the start of this epoch as 1784 with the beginning of the Industrial Revolution and the 'carbonification' of the atmosphere by the burning of coal.

If coal causes social, economic and medical trouble when it is dug out, it continues to pollute and endanger health when it is burnt. Sometimes the toxic events are dramatic as in Hangzhou, China, which was the location of an amazing incident. The country is used to many kinds of environmental disaster, but on 10 March 2013 the city was inundated by a kind of black rain that was infused with sticky coal dust. It covered every surface and remained for some time after the rain had stopped. For the most part, though, coal simply contaminates the air, causing smog, soot, acid rain and toxic air emissions. Despite the rise of petrochemical smog, coal is, according to the Union of Concerned Scientists, 'still the single biggest air polluter in the U.S.' (www.ucsusa.org). Burning coal releases chemicals into the air, including, nitrogen oxide, mercury, sulphur dioxide and particulate matter. Many governments have pledged to reduce this pollution by phasing out the use of coal, but around the world the consumption of coal continues to rise, often while states try to balance present prosperity with the plans for the future. For example, the Ruhr, in Germany, is one of the most heavily industrialized places on earth. When Mrs Merkel phased out the nuclear industry in 2011, coal consumption grew by 13 per cent. And Germany uses lignite coal in its boilers. This produces less energy and considerably more carbon emissions than other kinds of coal. In the same spirit, China is developing its wind and wave-powered technologies at a great rate, but remains a heavy user of coal. Tim Flannery blames both ignorance and political manoeuvring for the present predicament:

> Even in nations that lead in climate action, few politicians understand how dangerously and swiftly the burning of fossil fuels is altering our planet. Collectively, politicians are failing to maximize the chance of an acceptable outcome. A knowledge deficit among politicians is only partly responsible. Political lobbyists who set out to mislead are also to blame. (Flannery 2015: 81)

The concept of the Anthropocene does remind us that global solutions are needed to solve the problems of climate change and understand the function that our collective action plays in causing such damage to the atmosphere. If human activity has brought us to the point where we are destabilizing the entire planet, it seems that extraordinary effort will be needed to halt these destructive processes.

Photography has been recruited in the service of revealing the extent of pollution and dereliction. For example, in 2011 *National Geographic* magazine curated a slideshow under the title *Age of Man*. It was designed to illustrate the nature of the Anthropocene. Here were the oceans damaged by acids and pollution, the endless landscape of oil pumps, industrial agriculture, the vast detritus of industry and, of course, MTR coal plants (Bonneuil and Fressoz 2016). Some photographers have embraced the notion of the Anthropocene and attempted to look at the ways in which we now dominate nature and have become the most important geophysical force on the planet. In 1986 David T. Hanson's photographs of the coal-mining town of Colstrip, Montana, together with the scarred landscape that surrounded it, went on show at the Museum of Modern Art in New York. Colstrip is home to one of the largest strip mines in the United States and has developed into being a modern-day factory town. Hanson's colour photographs are formally and tightly composed, and this gives the mine and its surroundings a geometric precision that frames the devastated land. The skyline is broken up with pylons or factory chimneys. He records the spoil heaps, the bust-out trailers and the physical structures of the mine, including cooling towers and power plants. His images of the mine have a formal elegance, as do the fluorescent waste ponds that are abstracted into a kind of beauty.

Hanson's work is of first importance in exploring images of degraded landscapes, as is that of Richard Misrach, the distinguished photographer who has explored the American landscape for more than forty years. He was influential in using colour and large-format cameras and is often compared with the German artists Thomas Struth and Andreas Gursky. The landscape of the American West, especially its despoiled deserts, has been one of his preoccupations, but he has also recorded the degradation of the land brought about by human activities –

by tourism, the petrochemical industries and industrialization. He is interested in the social and political factors that lead to the ruin of the landscape, but he is also very attentive to the aesthetic qualities of his large, colour photographs. Art photography has become the dominant photographic form for exploring the Anthropocene. Gursky's colour images of enormous spaces, of skyscrapers, buildings and mountains, are often taken from a high vantage point and are awe-inspiring and disturbing. He reveals a world where nature has been abolished and everything is marked by human activity, by a restless, overwhelming power, a spirit that might be seen as being at the heart of the Anthropocene. Alix Ohlin has commented on Gursky's work in the following terms:

> The vertiginous dynamic of globalization, the subject of Gursky's work, is the contemporary locus of the sublime: a grand power in the face of which we feel our own smallness. (Ohlin 2002: 24)

However, a move to images that instil in us feelings of awe, and away from the production of formal pictures of beautiful or sublime landscapes, was not accomplished all at once. It owes much to a 1970s movement that was initiated by an exhibition in 1975/76. Called the *New Topographics: Photographs of a Man-Altered Landscape* it was held at the International Museum of Photography, George Eastman House, Rochester, New York. The show introduced work by Robert Adams, Lewis Baltz, Frank Gohlke and Stephen Shore. It also included photographs by the Bechers, whose work I discussed in Chapter 1. The exhibition moved to other venues and over time became extraordinarily influential, not only in the United States, but also in Europe. The Bechers' strict body of rules influenced the shape of the exhibition with each photographer able to show ten prints each 8" by 10 and, with the exception of the work of Stephen Shore, in black and white. While each of the artists became well known the New Topographics show was more than a collection of individual works because it proposed a new way of looking at landscape. The work of photographers such as Richard Misrach and Naoya Hatakeyama could scarcely be conceived of without its influence.

The celebration of the pristine scenery of nature now gave way to the examination of a landscape scarred by human activity, a landscape at once owned and neglected. It recorded what the French anthropologist Marc Augé was to call 'non places', which included gas stations, supermarkets, car parks, airports and banal and empty corners of cities (Augé 2008). Where street photographers were anxious to show the city as bustling with life, full of extraordinary encounters and random coincidences, the new topographics photographer revealed the city as a series of lonely tracts of real estate. There was no intention to add aesthetic value to these spaces. The use of the work 'topography', that is, the accurate and detailed delineation of a place, clearly described the intention of the exhibition. Nor was the work overtly critical of the society that created such flat and mundane spaces. It rather laid out a body of implicit rules by which landscape could be shown.

The most recorded sites were in the American West, a place steeped in myth and usually shown as harsh country softened by romantic blue hills and deep gorges with wide open spaces where once buffalo roamed. Now it was depicted as a place full of tract homes and dreary warehouses. The move away from romanticism, in favour of this banal aesthetic, did not mean that they adopted a documentary approach. The viewer was not invited to look at these sites and consider their social value, nor was there any photojournalistic interest in the places they recorded. The flat, stark and evenly lit images were impassive in the way that the Bechers had always been, and made no appeal to political action. They were, however, taken up by the conceptual art movement that was flourishing at the time. While that movement aimed to make art objects less important than the bodies of ideas that underpinned them, the new topographics movement stripped away from landscape photographs any journalistic or social element and made it available to 'art'. The body of ideas that underpinned the movement went on to be used in images of many other landscapes. It allowed artists to look afresh at the state of the land liberated from the baggage of inherited aesthetic ideas about beauty. It freed them to ignore the representation of 'scenery' in favour of looking at the used, despoiled and often broken structure of a landscape.

Burtynsky

Among the photographers who were influenced by the new topographics movement was the man who is surely the high priest of the representation of the Anthropocene, Canadian photographer and artist Edward Burtynsky. For decades he has been producing huge photographs of mines, quarries, oil fields and refineries. He has shown us the earth scarred, burnt, corrupted and ripped open. Strange rivers coloured scarlet, green or blue from the leakage of chemicals flow sluggishly through otherworldly landscapes, while luminous lakes glitter like strangely hued jewels in the bowl of otherwise deserted country.

Lori Pauli quotes Burtynsky as saying that he began this work because:

I had to cross some unknown territory in Pennsylvania, which happened to be one of the largest strip mining areas in the United States. All of a sudden I was in this town called Frackville and I thought, 'something seems different here'. I started to drive around the slag heaps and then finally stood in one spot. It was then that I realized that as far as my eye could see everything had been transformed, there was nothing natural left. Slag heaps and incredible aquamarine water. It was like another world. It was surreal. I thought, 'This is where I want to go. I want to do the mined landscape'. (Pauli 2003: 18)

Since then he has devoted much of his life to showing us the mined landscape. The sheer size of his images, often 100 × 150 cm, make it necessary for us to, as it were, enter the picture in order to observe the details and try to work out exactly what we are looking at. Viewed from above, from vantage points, cranes and helicopters, and often without a horizon line, the earth seems utterly strange, horribly damaged, but also peculiarly beautiful. We are appalled by the content of the picture, but seduced by the formal aesthetic of the piece, the crisp images, the saturated colour and the careful balance of elements within the frame. Burtynsky's work is often seen as an indictment of the big organizations that carry out this work, but he disclaims any intention of making 'political' interventions, and would not want to be regarded as a 'concerned photographer' in the spirit of, say, Sebastião Salgado. But, this is not to maintain that he sees his work purely in aesthetic terms, and as lying completely outside the social realm. In a speech he made when accepting a TED prize in 2005, while denying that he was 'against the Corporation', he said, 'I want to use my images to persuade millions of people to join in the global conversation on sustainability'.

Richard Baker has pointed out:

> Burtynsky's oeuvre pulses with questions about how we use the concept of art to neutralize the implicit moral or social force of photographic images. His pictures are unarguably striking and thoughtful enough to warrant description as art. But does appreciating, or merely accepting, photographs as art preclude being stirred to action by them for, say, a conservationist cause? (Diehl 2006: 40)

Writing about Burtynsky's subjects, Diehl calls them, 'Vast, unnatural terrains created by the machines, excavations and accumulated detritus of modern civilization' (Diehl 2006: 119). The article is in a review of Burtynsky's exhibition *Manufactured Landscapes*. It looks at the United States, Europe, China and India. Diehl suggests that the works 'emanate an overwhelming beauty' and carrying the look of 'old master paintings' such as Turner's *The Fighting Temeraire* (Diehl 2006: 130).

The sublime consistently served as a framing device for all Burtynsky's photographs. They evoke feelings of amazement and anxiety and his 2011/2012 exhibition was called *The Industrial Sublime*.

> The sublime plays a distinct role in making environments and environmental change visible, but it also conceals a certain visibility not born of the breathtaking, such as the particularity of detail in Burtynsky's work or the visually dull greying of everything by coal dust. (Schuster 2013: 194)

Burtynsky has defined his relationship to the making of landscape images. He said that 'I began by photographing the pristine landscape, but I felt I was born a hundred years too late to be searching for the sublime in nature' (Ottawa p. 47). Burtynsky's aerial views offer a synoptic take on the landscape, a grand sweep over the terrain. On closer inspection the photographs reveal the gashes in the earth. It is gouged, cut open, squeezed and slashed. Despite the intention of nations and communities around the world to dispense with coal and move to renewable energy, these are landscapes that dramatically reveal the overwhelming lust for coal and the dereliction that is tolerated in the search for it.

Even as this goes on, the history of coal, and of miners, is being celebrated in theme parks, museums, heritage trails and tourist mines. In the final chapter I look at these sites and examine what version of history they provide.

FIGURE 44 Peter Arkell, *South Hetton Colliery County Durham Closed Down June 1986 after End of the Miners Strike.*

7 HERITAGE, MEMORY AND NOSTALGIA

Coal and the heritage industry

Much of Western Europe has abandoned coal mining and it is in severe decline in the United States. The world of coal has been tidied up, cleansed and often obliterated. In 1975 Richard Llewellyn wrote a sequel to *How Green Was My Valley*. The second book praised the 'greening of the valleys' and applauded the fact that the rivers were running with fish and the hillsides blooming with newly planted trees (Llewellyn 1975). Sociologists, however, report that high levels of unemployment, physical and mental illness, depression and drug abuse mark the former mining valleys. However, in many countries, including Britain, there has been a concerted effort to reshape mining areas and to help generate local economies. For example, the Louvre has created a museum in Lens (a town in the coal mining area of the Pas-de-Calais) in order to bring visual culture to a depressed region, and to boost tourism both from within France and abroad. Opened by President Hollande in 2012, it attracted 900,000 visitors in the first year of its existence. In Britain the Coalfields Regeneration Trust was set up in 1999 to provide and support a range of community-based projects in skills development, health and education. It brokers partnerships and strives to lever funds from other bodies and businesses. However, its strategy document for 2016–2019 concludes with the message that the disadvantages coalfield communities still face in terms of employment, skills and health are clearly evident and the statistics support the ongoing need for economic and social regeneration.

In the 1980s coalmines were closing all around Europe and the United States. Often remote communities were left without an industrial base and once-thriving miners found themselves unemployed. Among the projects designed to ameliorate the condition of these places, and re-develop local economies, was the re-landscaping of pits and the establishment of what became known as 'industrial heritage sites'. These may incorporate a country park, a mountain bike trail, a hiking route or a chance to hunt or fish, but at the heart of them is a colliery, presented either as ruins or as a 'mining experience'. There are trips down coal mines, where with a hard hat and Hi Vis jacket, one is led by an ex-miner through some clean, uncluttered, underground tunnels. The economic rationale for such places has been expressed by J. Arwel Edwards:

> As a result of the closure of the mines, communities suffer a process of economic, social and demographic crisis and attrition. This situation forces the local authorities to search for alternative forms of development and economic reactivation on the basis of the irreversibility of the disappearance of the mining monopoly. All the projects seek a solution through the diversification of the economy and, in this attempt, a significant role is played by tourism promotion projects, based on their mining, historic and, in several cases, environmental heritage, thus combining industrial manufacturing and natural elements. (Edwards and iCoit 1996: 354)

Today, throughout the world, in what was once coal-mining country, there are hundreds of such places. UNESCO lists among many others those in Chile; Nord-Pas-de-Calais, France; an industrial complex in Essen, Germany; together with the Zollern Colliery in the transformed Ruhrgebiet; an old coal-mining town in Sumatra; and Big Pit Blaenavon in Wales. There's a coal-mining Heritage Association in McCoy, Virginia, USA, and a Coal Mining Heritage Park and loop trail in the same state. Kentucky, Illinois and Pennsylvania have many examples of the same kind. In Australia one can visit a Coal Mining Historic Site on the Tasman Peninsula, which tells the story of the employment of convicts in early mining. Some eighty-two coal-mining sites in New South Wales are listed on the Australian National Trust Site. In addition to the outdoor sites there are mining museums in thirty US states; in Estonia, Cape Breton, Wallonia, England, Scotland and Wales; and in many other places. In short, what were once paradigmatic places of labour have been converted into sites of leisure.

Industrial tourism

The history of tourism has been a story of the extension of places that are, for one reason or another, and for different reasons from time to time, confirmed as locations that we should aspire to visit. In this sense, it makes special many sites that have had otherwise a perfectly mundane existence. From the 1970s a new museology began to articulate ways in which the old industries could be presented to visitors. In France from the 1970s ecomuseums began to put together the museum with ecology. Building on the existing open-air museums, they attempted to work with local communities in order to protect the physical and cultural resources of a region (Howard 2002). Such museums were soon to be found in other parts of Europe and in China. They also influenced the industrial heritage museums of Britain as well as the neighbourhood museums of the United States. The development of these heritage centres has raised questions about their ability to regenerate the area in economic terms, but also about the authenticity of the tourist experience and the nature of the nostalgia that draws them there. By the end of the twentieth century, heritage had become central to economic development in mining regions. No longer seen as a marginal activity of archivists and popular historians it was a concept that had to be factored in to any economic redevelopment plans. In the 1980s, however, tourism of all kinds was subjected to a great deal of critical attention. The key points of debate were the relationship between the terms 'heritage' and 'history' and the subjective experience visitors might have in heritage places. Historians feared that nostalgia would be the dominant emotion and would militate against a rational study of the past. Part of the critique came from theorists of the postmodern, who contended that we were living in an unstable world: one that was out of touch with the material and the authentic. It was argued that we were no longer grounded in a world governed by 'facts', but existed in a semiotic system of free-floating signs. In these accounts, tourism, with its dislocations, is a paradigmatic experience of the modern (or postmodern) consciousness. At the same time, some historians regarded with critical suspicion the idea of 'heritage', which smacked of a nostalgic longing for a safe, rich, settled past. Industrial archaeologists, who were knowledgeable about machines and technical processes, looked on the new tourism as a leisure activity that lacked serious intent or heuristic value. Yet many of the sites were popular and well attended. Their interpretation centres with photographs, audiotapes and film often gave a good impression of the nature of industrial work and life. In an important survey Rosemary Power found that local communities largely welcomed the new sites:

> Heritage was seen as valuable for areas seeking to retain pride, identity and the means for regeneration, in an increasingly mobile society. There was also a strong concern that many of the social and community elements seen as intrinsic to the to the heritage of coalfields' communities were no longer available to society at large. (Power 2008: 168)

So, while communities drew strength from the industrial achievements of the past, it seemed that the old values may have deserted the living communities and could be found only in museums.

Heritage tourism and modernity

This loss of the values engendered by an industry was one of the ways in which it seemed modern society was changing profoundly. In his book *The Tourist*, Dean MacCannell tells us that wherever 'industrial society is transformed into modern society, "work" is simultaneously transformed into an object of touristic curiosity'. The argument, then, is that in a post-industrial world, 'work' has been altered into spectacle, and our own subjective condition is that of the spectator. For

MacCannell, modernity is grounded in instability and inauthenticity; industrial culture is increasingly undermined by 'alienation' and 'affirmation of basic social values is departing the world of work and seeking refuge in the realms of leisure' (MacCannell 1976: 6).

According to these ideas the decline of industrial culture has led to the 'end of work' and, in some instances, labour itself has been transmuted into a tourist experience. Of course, the idea that work might one day be written out of human experience has a long history, especially in utopian writing, and has received new attention with the rise of robotics and digital technologies.

In 1980 Ándre Gorz published his very influential book, *Farewell to The Working Class*, a work that attempted to imagine the future when human toil had been taken over by automatons. He argued that:

> A society based on mass unemployment is coming into being before our eyes. It consists of a growing mass of the permanently unemployed on one hand, an aristocracy of tenured workers on the other, and, between them, a proletariat of temporary workers carrying out the least skilled and most unpleasant types of work. (Gorz 1980: 3)

This dystopian view of future patterns of employment may have seemed exaggerated at the time, but almost forty years later it appears to be a possible trajectory along which we are moving. Certainly, some kinds of work that were seen as arduous, unattractive or mundane have become objects of mass contemplation – of tourist attention. This might be seen simply as a consequence of the democratization of leisure and travel. Just as in the eighteenth century ladies and gentlemen journeyed to look upon nature and examine the peasants, and in the nineteenth century the bourgeoisie toured factories and the sites of great engineering ventures, so in the twenty-first century all kinds of people take trips to look at places of abandoned industry. To the castles, palaces and formal gardens that were once the central sites of 'our heritage' have been added redundant steel works, early beam engines, quarries, mills, shipyards, factories and, of course, coal mines.

Tourism and photography

If tourism is regarded as an inauthentic form of existence, photography was seen as the quintessential technology for constructing and distributing the world of ersatz signs within which we were said to exist. Certainly, photography is an essential feature of modern tourism. An older conception of the camera as an unblinking eye that revealed 'facts' was replaced in postmodern accounts by the notion that photography, through the endless production and circulation of images, created a free-floating and volatile world of signs. Moreover, the photograph, and later films and videos, were also seen as replacing direct contact with the world. David E. Nye gives a striking example of this:

But increasingly advanced technologies not only get people to the site; they also provide alternatives to seeing it. A maximum of 92 people a night are allowed to stay at the Phantom Ranch at the bottom of the Grand Canyon: but outside the park 525 people per hour can sit in an IMAX theater and watch *The Grand Canyon – The Hidden Secrets*, a 34-minute film shown on a 70-foot screen with six-track Dolby sound. As an additional attraction, the theater advertises 'Native Americans in traditional dress on the staff'. During the performance guests are encouraged to 'enjoy our fast food, popcorn, ice cream' and other snacks. Why bother to hike into the canyon when all the highlights have been pre-packaged? (Nye 1994: 289)

The idea that a real experience may be replaced by a visualization of a place is one way in which the audio-visual functions. Today, however, the camera phone allows for a stream of images in which the whole of our everyday life may be recorded and modestly celebrated. Famous sites now need to be authenticated not by possession of a picture in an album, or by watching a movie of them, but by an image of us standing and smiling before them. But photographs always validated our experience of 'being there', which is not merely one of visiting an unfamiliar place, but of claiming the authentic experience *of* a strange place. Photographs are records and documents that pin down the changing world of appearances. Now we really want to put ourselves in the picture, and the selfie merges us immediately and seamlessly with the place in which we stand. From the nineteenth century, photography also played a part in recording not the rise of industrial tourism, but the decline of industry itself. Established in 1908 the Royal Commission on the Historical Monuments of England was set up to record buildings and monuments of historical importance, and an archive of these was established. In its 1985 publication, *Industry and the Camera*, the Commission pointed out that 'in an historical perspective of millennia it seems the very essence of industrial development that individual industries fall even more surely than they rise' (HMSO 1985: 3). It continues:

As a consequence of this inevitable decline, abandoned mines, factories and shipyards provide 'an archaeology of industry represented by redundant plant and other remains 'and these need to be sought out, photographed, classified, labelled and stored in suitable archives: As the national economy changes from one to a large extent based upon heavy industry, the need and opportunities to record both the processes of change and their consequences in terms of relict plant and even whole landscapes fall firmly within the remit of this commission. (HMSO 1985: 6)

So, even while the miners' strike was still raging, the Commission had hauled coal mining, and the landscape within which it was set, into its orbit. It turned heavy industry into a monument of the past.

Its specific purpose was to encourage anyone 'with a good camera' to undertake the task of recording the vanishing face of this industrial world. Photographs, the report says, constitute a major resource for the historian, and *Industry and the Camera* included a selection of photographs designed both to illustrate the industries of the past and to act as examples for those who wanted to record contemporary dereliction. The book drew attention to the neglect of industrial subjects by photographers in the nineteenth century. Indeed, in early photography, there is a scarcity of images of workers, except where they are gratuitously caught in the shot or placed so as to give scale to some mighty engineering project. Thus, while photography told us a good deal about technical processes and infrastructures, it offered little useful knowledge about the world of work viewed as a human activity. Again, it found it hard to represent collective action or explore the world of workers' organizations and patterns of response to industrial labour. To these absences we need to add the fact that photography foregrounded particular kinds of work at the expense of others. While heavy industry was always fascinating, other trades and labour were ignored. Domestic labour, for example, lacked the machinery and interesting equipment through which the drudgery of a large section of the workforce might have been displayed. The same is true of the army of clerks and office workers that was so central to Victorian society, yet has passed without any real visual record. While the urban poor, the down-and-outs and the dispossessed were always subjects of great interest, not least because they were visibly on the street, the average worker went largely unrecorded.

The Royal Commission's report was produced at a time when the strategy of the British Government was to let industry fight for its place in the market and, if it declined, to allow the 'service sector' to replace it within the economy. The new tourist sites, which were in part produced by this policy, included many which took the abandoned industrial works as the objects of their attention.

Such sites could not exist without the interpretative aid of photography. But, as with the photographs of the Royal Commission, the world of the early industrial museums was one in which work was largely revealed to us through machinery, processes and technologies. The industrial sites did not initially offer a study of human labour so much as an examination of how labour had been structured, disciplined and controlled so as to service particular technologies. Because we were placed in the past, often without any clear definition as to the exact period, these technologies emerged as neutral and natural. The tour guides at mining sites often stressed the onerous or degrading nature of the tasks performed by workers, but it was difficult to connect them into a developed narrative of workers' lives. As the tourist trade became more sophisticated, the tourist sites used more and more family snapshots, oral histories, accounts of working lives from the people that lived them, together with archive film so that the subjective experience of workers became much more important.

Nostalgia

The argument that the industrial sites offer a nostalgic peepshow into a largely fictitious past is one that runs through a whole strand of writing on tourism. For example, Robert Hewison contended that the 'heritage industry' is to be deplored because it 'draws a screen between ourselves and our true past' (Hewison 1987: 10). Moreover, he argued that an obsession with the past distorts our sense of ourselves and our own contemporary culture. This notion of a 'true past' is interesting, especially as Hewison sees 'heritage' as a cultural form that is generated and validated by nostalgia. This is a not unusual double critique of such sites: that there is a 'real' history which they fail to represent, and that they engender in the tourist an inappropriate response to the past: nostalgia rather than detached consideration. We may think of this as a simple nostalgia, but there are more complex forms of this condition and it is worth looking at the origins of the concept of nostalgia.

The word entered the French dictionary in the seventeenth century and was widely used as a clinical term for the next 150 years. Unlike vague feelings of melancholia or homesickness, it was regarded as a serious condition that might, in extreme cases, lead to death. It began as a yearning for home by soldiers posted far away, but it took on a temporal condition when the longing that it engendered was not for a particular place, but for the past itself: a feeling so strong that living in the present became insupportable. This excessive attachment to the past was connected to the craving for home, family or the place of one's birth. So there was, from the first, a close connection between time and space that has never left our conception of nostalgia, even if, from the middle of the nineteenth century, it was no longer seen as a clinical condition (although it regained this status for a while in the American Civil War). Gradually, it began to be looked on as an illegitimate emotion, one that allowed the sufferer to luxuriate in some notion of the past instead of coming to terms with the present. In its original form nostalgia is held to have died out because modernity abolished the special nature of individual places. In a fascinating article on nostalgia, Michael Roth quotes the French physician Pilet, writing in 1844:

> One has to say it, nostalgia is a disease that is tending to disappear and that is observed less frequently each day: the establishment of rapid communications, the inundation of a ceaselessly invasive civilization, erases one by one the moral colours that so often created the disease in other times. From the general cosmopolitanism we see that in becoming attached to everything one is no longer attached to anything ... the further we move away from a state of simplicity, the less we cling to the tombs of our ancestors and to the soil on which we were born. (Roth 1992: 278)

Nostalgia, then, was dying just as photography was being born. Indeed, the very factors that made people doubt whether photography was an art – its contingency, its inability to select from what was in front of the lens and its consequent richness of ethnographic detail; its melancholy inscription of the passage of time within its images – all gave the new medium a curious relationship to the past. Over time the emotion of nostalgia could lose its inchoate quality and be evoked by a photograph itself. From its inception, photography was an instrument for surveying, classifying and documenting aspects of social life, and it has become one of the key forms through which the past has been recorded and displayed. Indeed, we could no longer consider the Victorian question of what constitutes a *proper* memory without looking at the way in which photographic images have shaped our personal, social and collective views of the past. Generalized photographs of the past surround us in pubs and coffee shops, heritage sites and advertisements. Black and white or sepia toned, they make an implicit claim to authenticity, tradition and a respect for the past, while bright colour images evoke modernity, vibrancy and the future.

Nostalgia for coal

Quite frequently people are nostalgic not for their native village or the years of their childhood, but for a social life that they have never lived, a 'past times' that seems richer, more amiable or more truly authentic than anything that can be found in our present existence. Nostalgia for coal mining is, naturally, strongest in former mining areas where it summons up a sense of warm communities, financial security and honest labour, but it can also be found among people whose connections with the industry date back several generations, or are non-existent. Many people live in a world of no or low-paid work; millions more in a place where work is abstract and immaterial. The coalface that office workers and managers claim to have slogged away at all day turns out to be a computer screen or a mobile phone. Under these conditions the life of a fisherman, a steel worker or a collier is certainly not envied, but there is a nostalgic sense that work of this kind ought to be going on all around us and is somehow more real than our present toil.

For people living in mining districts, nostalgia is a major emotion because it is one of the ways in which community memory is kept alive and a sense of the past is engendered and sustained. There are literally many thousands of books of photographs of mining in countries, regions and districts. In many Western European countries the landscape of mining has been taken back to some notion of the rural past, so that the markers of industrial work have been obliterated. What remains are portable artefacts and photographs.

Souvenirs

Souvenirs of mining, often described as 'collectables' would appear to be big business. In Chapter 1 we noted Peter Fuller's claim that nothing can be made

FIGURE 45 Mining Souvenir, Hard Gannin (Hard Going) when it was difficult for a pony to move the tubs themselves the miners had to play a part.

from coal. In fact many objects are carved from anthracite or from a resin designed to look like coal. It is possible to acquire a vintage French letter rack decorated with the head of a coal miner and a lamp in copper. You can buy coal figurines online or from heritage shops; eBay is awash with them. The Coal Mining Museum of Slovenia offers an extensive range, as does the gift shop at the Lackawanna Coal Mine Tour, Pennsylvania, and many other places. Collectables bought at these sites may function as simple reminders of a day out, but, unlike the photos on mobile phones, they record not the trip itself, but are souvenirs of the coming together of the present with the imagined past. Here are statuettes of miners in tin baths, underground rescue workers complete with breathing equipment, standing figures with lamp and shovel, and a lad leading a pit pony. Susan Stewart argues that the 'capacity of objects to serve as traces of authentic experience is, in fact, exemplified by the souvenir'. She goes on to observe that:

> The souvenir speaks to a context of origin through a language of longing, for it is not an object arising out of need or use value; it is an object arising out of the necessarily insatiable demands of nostalgia. (Stewart 1993: 133)

Donald Trump's pledge at the start of his presidency to 'bring back coal' was seen by many critics as playing on a 'nostalgia for coal' that connected the president with working class voters who still long for work, wages and the culture of heavy labour. The industrial heritage industry would seem to show that many people have at least a passing interest in mining sites. Noting that cultural heritage has become one of the ways in which the coal industry permeates life in Australia, Pearse et al. describe the Blackwater International Coal Centre (BICC) which is sponsored by a number of coal companies:

> It's a museum and open-cut coal mine tour service rolled into one, designed to 'provide a sophisticated and enduring platform for showcasing the mighty Australian Coal Industry and the associated industries that underpin the state and federal economies'. Packaging coal mining as a fun heritage attraction, the BICC is also perhaps as close as it gets to a coal based theme park. (Pearse et al. 2013: 123)

In France The Mining History Centre is a large mining museum. It is located on the site of a disused colliery at Lewarde in the Nord department. More than 150,000 people visit it every year. So, despite all the many problems that coal generates from the act of digging it up, to cleaning, washing and transporting it, to the noxious carcinogenic smoke it emits when being burnt, many people enjoy the experience of being a visitor at a coal museum or tour of a mine.

New kinds of journeys are undertaken to get in touch with the vanished workers, and without the embarrassment of those 1930s writers who tried to identify with the proletariat. For, in the democracy of the turnstile, we are all equal before these displays. We no longer need to tremble at the thought that we might be found wanting, for we are being guided by that most important element of the experience – a real ex-miner who knows his way around the site and is the guarantor of the veracity of our experience. That urgent cry of the documentaries, 'this is how it is', has given way to the calmly nostalgic view, 'this is how it was'. But when? Most sites are not grounded in a specific, historical moment. More importantly, the dynamic pace of change, the restless energy of capitalism, produced workers who moved to the collieries in the hope of material improvement and moved elsewhere when times got hard. Because artefacts have the enduring properties of iron and steel, they appear solid and neutrally authentic. But the introduction of artefacts, when they were clustered together as technologies, was often a matter of bitter contest, conflict and dispute. Each major technological change brought about new ways of working which, in turn, changed the relations of production and the relationships between one group of workers and another within and outside the pit. They brought with them new kinds of supervision, new disciplines and new ways of calculating wages, which were resisted often with a considerable struggle. The much-praised 'unique communities' were formed in part out of such struggles. They were unique precisely because 'industrial disputes' were not confined to the point of production, but called on and shaped the wider social world. It is clearly impossible to get any sense of this from looking at the machinery and organization of a coalmine. We are not dealing with some simple divide between past and present in these places, but with a space that is complexly articulated by several, competing 'times'. We must remember that tourism is not simply the product of a desire to visit unusual places; it is also the result of careful planning by state and regional organizations, who see 'heritage' as a way of regenerating former industrial areas.

Memorabilia

In 1985 John Gorman published a selection of memorabilia from the National Museum of Labour History in London. There were studio photographs of significant figures in the Labour movement as well as pictures of women and children working in sweat shops or 'homeworking' with the whole extended family making simple products around the kitchen table. More dramatically were the photographs of a hunger marcher being arrested, or tenants taking part in a rent strike. But the book also included posters for a range of national and international causes, paintings, drawings, leaflets, ceramics, badges and banners from organizations such as the miners' protection society (Gorman 1985). In Britain, trade unions had gained a limited freedom to function when the Combination Act was repealed in 1824. Homemade banners emerged at this time and had become very popular by the end of the nineteenth century. At that time commercial banners were introduced and were used in numbers of trades and industries. The suffragettes carried them, as did temperance groups and churches. Mining lodges, union areas and even individual pits commissioned banners that could be unfurled on marches, demonstrations and gala days. They were present during the strike of 1984–1985, and many miners marched back to work at the end of the strike with the banners held high. They were most common in the north of England, but existed elsewhere. The Bargoed NUM banner showed a prosperous looking couple with two children walking through sunlit fields with the slogan 'Peace and Prosperity'. This is a common example of the iconography of banners, but more complex versions were often inscribed with pit head wheels, masonic symbols, Old Testament scenes, Greek letters and two clasped hands.

There were also portraits of union leaders and Stakhanovite images of heroic miners. The premier commercial maker of mining banners was the firm of George Tutill which produced oil-painted, double-sided silk banners which became the technical standard for superior banners. Often subtle and complex the banners, carried aloft by two people, are regarded as fine examples of working class art. The death of the mining industry means that banners are no longer being commissioned, but they are seen as an important resource for historians, so the search for neglected and misplaced banners is an active one.

FIGURE 46 Dean and Chapter Banner. Original in colour.

Banners are important objects in the mining museums and heritage sites where they are displayed both as part of the narrative of the history of mining and as cultural objects in their own right.

The Durham Miners' Gala was a major event at which many banners would be unfurled. The death of the industry has not led to the end of this celebrated event. It continues, grows in attendance and is still a great annual occasion for carrying banners. Indeed, the banner has become symbolic of the qualities of the old mining communities. Chris Scott tells us that even where a colliery is intact, although used for different functions, 'these are not treated or considered with anywhere near the same interest or reverence afforded to a banner' (Scott 2009: 70). It seems that in the colourful, beautifully sewn, silk banner, the nostalgia for coal has found its most potent symbol. Scott recognizes the importance of the banner as an emblematic object, but also points out that 'At least for sections of the community, it is essential that the banner is displayed in the locality, in order that the sense of folk memory and the local distinctiveness it creates are firmly rooted' (Scott 2009: 71).

New banners are being created that often memorialize the past and refer to times when the industry was buoyant, and some heritage sites have created groups to help in the making and using of banners. It seems that an interest in banners is a gateway to the study of mining heritage. This goes together with a renewed interest in the collection of oral histories and the production of memoirs from former miners.

Mining memorials

The melancholy engendered by the end of a particular way of life is mirrored in a new interest in the physical relics that record and memorialize mining. The Welsh village of Senghenydd is sadly famous as the home of an explosion in the colliery in 1913 that killed 439 miners and a member of the rescue brigade. The artist Les Johnson was commissioned to mark the centenary of the event (which is still well known in Wales, as it has been often rehearsed in popular memory) to create a bronze sculpture. Entitled *The Rescue*, it shows a man holding up a miner's lamp and supporting a figure that is nearly collapsing. Located in the memorial garden, it is one of many artworks that are designed to keep and celebrate the mining industry. Often these do mark the sites of serious disasters. There is one at Six Bells near Abertillary. Called *Guardian* it is the work of Sebastien Boyeson and was completed in 2010, the fiftieth anniversary of the 1960 disaster that killed forty-five men at Six Bells colliery. A figure, some 20 metres high, stands on a sandstone plinth and dominates the site of the now-defunct pit. Most of these memorials are figurative and cast in bronze or a resin designed to look like bronze. A standing male figure on a plinth holding an implement inevitably conjures up war memorials, hundreds of which were created after the First World War, so that almost every town in Britain, France or Belgium has one – a soldier standing proudly or leaning on his rifle with the names of the fallen dead inscribed on the plinth. Mining memorials that are sited on disaster sites often repeat this exactly. As examples of public art they are modest and conventional, but it seems that there is a general desire not to let the past slip away without some public recognition. So, in a casual and uncoordinated way, miners are being inscribed within a formal system of commemoration and gaining the status of the fallen wartime dead.

There are plenty of other places where memorials to miners can be found: The Richlands Coal Miners' Memorial in Virginia has a bronze miner strolling out of a memorial building. Behind him on the wall is written, 'In Honor of Our Coal Miners'. In Milau, France, a quite different style memorializes mining. It consists of an enormous gas testing lamp. Made of stainless steel and brass, it towers above the houses. But in McAlester, Oklahoma, the statue to the 'pioneer miner' erected in 1992 is, once more, a miner standing on a plinth. There's one in Paris, Arkansas, and another in Donetsk, Ukraine. Together with many, many more in Britain and in mainland Europe, Canada and Australia.

Of course, in the usual paradox of mining, there are still many deaths in the industry around the world. And these tend to go unremembered and without memorial. But, despite the continuing high death toll, there are some monuments to miners. For example, in China a significant one is at Fuxin in Liaoning Province.

FIGURE 47 *Universal Colliery, Monument to Coal, 1996*. One of a series of coal monuments by Paul Cabuts. Original in colour.

It commemorates the miners who died when forced to work underground for the Japanese during wartime. It is estimated that hundreds of thousands of miners died in this work. The monuments are sometimes to honour specific, often named, dead. At other times it is mining and the spirit of the industry that is being celebrated. Bargoed has a huge piece of public art. Designed by Malcolm Robertson, it consists of three miners' heads, standing almost 4 metres tall. They stare out at the top of the town. Inscribed on the steps leading up to the heads is written:

> With pride remember them for us a future they built, remember the miners. Our town slowly built around the mine, developed into something fine.

A typical mining community now needs to be reminded of its origins and the centrality of the industry to its history

Ruins

Not all memorials to mining are commemorative statues or sculptures. Some are simply disused mines that are kept in a safe condition so that they may be visited.

There are dozens of these in Pas-de-Calais alone and visitors can look at miner's cottages, mine buildings and slag heaps. More extraordinary are huge projects that consist simply of abandoned mines. It seems that there is an appetite for visiting all kind of mining sites. On 28 April 2017 the *Daily Telegraph* reported that CNN had listed the ruins of Cornish tin mines as one of the 'must see places before you die'. Often perched at the edge of the sea and blending into the landscape, the abandoned wheelhouses and pumping engines are listed as a UNESCO World Heritage Site. Now they are to be seen as examples of beautiful dereliction and may be found on innumerable tourist postcards. The ruins of coalmines are also of increasing interest, but many are far from picturesque. It's possible to categorize these ruins in a number of types: those that are simply abandoned and often create physical and environmental damage; those that have been re-landscaped and returned to a state of 'nature'; and those that have been incorporated into interpretive and heritage sites.

Abandoned sites of mining are common around the world. The marks of mining are inscribed on the landscape, sometimes for decades after the coal has been extracted and the pit fallen silent. Land subsidence, unstable slag heaps and disused mineshafts are common features of the environment in many places around the world, as they have been for centuries. When, in *Hard Times*, Dickens needed to have Stephen Blackpool disappear and so apparently break a promise, he simply had him fall down a disused mineshaft and not be discovered for several days (Dickens 1854). Clearly, his contemporary readers didn't see this as an unlikely accident, nor would people in India or Africa today. But these are far from the only ugly legacies of mined land. Greenpeace lists many environmental hazards associated with mining, including contaminated water supplies, the release of methane into the atmosphere, a lowering of the water table, long-lasting underground fires and acid mine drainage. But people still struggle to get at supplies of coal. Sometimes conventional mines that have been abandoned because they are uneconomic are simply re-inhabited by bootleg miners who work the easy seams for small amounts of coal.

But pits do fall into disuse and become abandoned wrecks. Ruins have long been regarded as a central feature of picturesque scenes, and, in the romantic representation of ruins, photographers took up where etchers and water colourists left off. There is, however, a particular problem with the ruins of coalmines, for what is left on the surface is only the superficial aspect of the hidden mine itself. Sometimes what we see are abandoned lamp rooms with lamp checks that will never be collected, hanging on rusty hooks, and melancholy signs whose advice and prohibitions will never now be heeded. There are many complex ruins of coal mines. For example, in Australia it is possible to wander among the fallen buildings and ruined walls of the *Coal Mines Historic Site* where once convicts were sent when sentenced to a period of hard labour. Only the outline of a chapel, store and bakery can now be seen as the whole thing has been gently decaying since it was closed in 1848. The website for a mine on the island of Ikeshima, a day trip from Nagasaki, Japan, specifically notes that 'ruin fanatics' will have an equally fine time as cavers and family groups. In his collection *Modern Ruins*, Shawn O'Boyle photographed deserted coalmines together with steel plants, prisons and mental health facilities (O'Boyle 2010).

Other places have taken on a romantic air, a fine example of which is Pyramiden, the former Soviet mining settlement located deep in the Arctic Circle on Norway's Svalbard Archipelago. Once regarded as a plum posting for Russian miners, now it is a ghost town with very many buildings. It gives us a clear view of how things were until 1998 when the mine was closed and its inhabitants returned to Russia. Only six people remain, one of whom, Aleksander Romanovsky, is the subject of David Beazley's short film, *Pyramiden*, which takes us into some of the remaining rooms. This was a town with many facilities, a gym, a swimming pool, a library, an auditorium for shows and movies and so on. Now deserted, photographs show a statue of Lenin gazing out to some large slag heaps. Animals increasingly populate Pyramiden; polar bears wander in, as do arctic foxes and reindeer. Also dropping in are tourists who are guided by Aleksander through the crumbling, echoing halls and corridors to empty rooms and drained swimming pools.

Hashima (Gunkanjima) in Japan is an extremely dense city situated on a small island. It once seemed to have had a collectivist ideology and presents depersonalized architecture in a brutal and rationalized style. The photographers Yves Marchand and Romain Meffre created a very powerful book on the ruins of Detroit, after which they turned their attention to the very different site of Gunkanjima. They see the ruins as an inevitable part of the project of modernity:

FIGURE 48 Manfred Thürig, view of ruined buildings on Pyramiden, 2015.

Gunkanjima … seems to be the ultimate expression of the relationship between architecture, culture of labour and the principle of industrial modernity, which not only aims at innovation and growth, but also at the effacement of any obsolete form. (Marchand and Meffre 2013: 9)

Hashima (the island on the edge) is an uninhabited island located some 15 kilometres from Nagasaki. Fifty years ago it was home to a very prosperous coal mine with tunnels running out under the sea. It also contained a thriving community with very high density of population. We need to remember that Japan was outside the Industrial Revolution until 1853 when the arrival of the US fleet, together with representations from Britain and Russia, opened the country up to international trade. In 1868 British mining engineers were employed to drive a vertical shaft mine at Takashima. They found a viable coal bed and demand for this good-quality coal soon grew. This led to a rush to open more mines including one at Hashima. The Mitsubishi company obtained the rights to the coal mines and the shipyard at Nagasaki. In 1890 it bought the whole island of Hashima and expanded production. It used the waste slag to build up the island. It constructed rows of wooden dormitories to house the miners, many of whom were convicts sentenced to hard labour. Military victories against both China and Russia in 1895 and 1905 allowed Japan to enter the league of industrialized nations. At Hashima they constructed a high concrete wall around the island, giving the place the uncanny appearance of a battleship riding the waves. Because of this physical appearance a journalist named it Gunkanjima (battleship), and this was how it became known in common parlance.

The island was producing around 150,000 tons of coal annually, and some 3,000 people lived there when in 1916 Mitsubishi built a reinforced concrete housing block, the first concrete building of any size in Japan; more housing followed, until the island contained thirty-one concrete apartment blocks. Coal production reached 410,000 tons in 1941, but it was won at great cost to the workers who were pent up in the grim blocks with little free time or leisure facilities. The conscription of Japanese workers into the army in the Second World War led to Chinese and Korean nationals being forced into working underground. By 1945 more than 1,300 labourers had died on the island from pit accidents, illnesses associated with exhaustion and malnutrition, and suicide. In the post-war reconstruction the coal of Gunkanjima became very important. Workers flocked there with their families for the first time in the island's history. By 1959 some 5,300 people were jammed into the housing blocks. It was said to be the highest density of people ever recorded in the world. By the 1960s there were schools, a gymnasium, cinema, temples, shops, a hospital and so on, but only two telephones.

In many ways working there was better than toiling in coalmines in the dingy towns of Kyushu and Hokkaido, because of the paternal influence of Mitsubishi. Apart from coal, everything had to be imported. There was no soil and no vegetation. The colliery went downhill in the late 1960s when oil began to replace

FIGURE 49 Gunkanjima, 'Battleship Island', Nagasaki Prefecture, Japan. Photographed by Kamal Parsi-Pour.

coal, and in 1974 the closure of the mine was announced. Since then it has been battered by typhoons, with huge waves washing away wooden structures. Visitors can now view it from a safe distance and it hopes to gain UNESCO heritage status. The beauty and symbolism of Gunkanjima lie in its decay. Photographs give us a view of the place from its site in the sea to small details of abandoned tools, musical instruments, radiators or television sets. In their book Marchand and Meffre also reproduce old black-and-white photos of people using the many facilities of the town next to an image of what has happened to it. Thus, a smart and busy barber's shop now lies in ruins, its padded chairs eaten away, its walls destroyed and the floor covered with detritus. We can see the appeal of these places as being one of re-imagining the past through the texture of walls and materials, through the echoing emptiness of huge spaces and the quiet of rooms that must have once resounded to voices and music. But, as Tim Edensor reminds us,

> [Yet] ruins do not merely evoke the past. They contain a still and seemingly quiescent present and they also suggest forebodings, pointing to future erasure and subsequently, the reproduction of space, thus conveying a sense of the transience of all spaces. (Edensor 2005: 125)

The 'transience of all spaces' is particularly marked in coal mining, for no mine goes on forever, and the closure of worked-out pits has always occurred. But industrial ruins hint that physical work itself may be transient, and disappearing with it would be entire cultures.

CODA

Most people think that we must give up using coal to save the planet from its awful pollution. Some dedicated ecologists even see nuclear fuel as a more acceptable option. George Monbiot argues that:

> The tiny risk imposed by nuclear power has both obscured and invoked the far greater risk imposed by coal. Scare stories about nuclear power are a gift to the coal industry. When these stories are taken seriously by politicians – as they have been in Japan – causing a switch from nuclear to coal, they kill people. (Monbiot 2017: 170)

Some groups try to hold on to the idea that coal has a real future. They advertise the virtues of 'clean coal', which may be obtained through elaborate carbon capture and storage schemes and expensive scrubbing and washing technologies, but most analysts of the industry see this as prohibitively expensive.

Meanwhile dirty coal is still a major fuel and produces some 40 per cent of the world's electricity.

In North America and Western Europe there are very few miners left at work and their numbers decline every year.

In the United States because of the election strategy of Donald Trump miners have become much discussed figures who are seen as symbolic of the working class, or the underclass. His Democrat opponent also spent some time discussing coal. Hillary Clinton said:

> I'm the only candidate which [sic] has a policy about how to bring economic opportunity – using clean, renewable energy as the key – into coal country, because we are going to put a lot of coal miners and coal companies out of business. Hillary Clinton (Speech in Columbus, Ohio, March 2016)

Commenting on this remark in her 2017 book, *What Happened*, she wrote:

> I made this unfortunate comment about coal miners at a town hall in Columbus just two days before the Ohio primary. You say millions of words in a campaign and you do your best to be clear and accurate. Sometimes it just comes out wrong. It wasn't the first time that happened during the 2016 election, and it wouldn't be the last. But it is the one I regret the most. The point I had wanted to make was the exact opposite of how it came out. (Clinton 2017: 263)

Clinton may well have had a fully worked-out economic plan for the regeneration of the coal regions, but what she did not do was subscribe to the idea of coal culture. She failed to tap into the imaginary of coal, to that place in the past full of well-paid jobs located in warm-hearted, friendly and supportive communities. She failed to subscribe to the notion of a place that valued fine family values, neighbourly support and strong, independent, folk. Later she commented on the centrality of the miner to working class life:

> For many people, coal miners were symbols of something larger: a vision of a hardworking, God-fearing, flag waving, blue-collar white America that felt like it was slipping away. If I didn't respect coal miners, the implication was that I didn't respect working class people generally or at least not working-class white men in small towns and red states. (Clinton 2017: 264)

Such communities are, indeed, slipping away, and we all know that coal is doomed. But it is taking a very long time to die. Today, all over the earth, people are still mining coal in all the ways we have looked at in this book. They are dredging small lumps from the sea or scratching them from the earth. They are cutting into the sides of hills and bringing out coal with ponies or mules. They are descending thousands of feet in cages to run machinery that will undercut the seams and allow the coal to tumble onto a moving belt. Others are hacking away by hand at the coal face as their fathers and grandfathers did. They are driving massive machines that tear open the earth in order to scoop out the coal.

But the only promising future in the coal industry seems to be to become part of the heritage movement – to visit a tourist mine and celebrate coal and the unique culture it created in a museum.

REFERENCES

Actionaid (2014), *South African Coal Mining 2014: Dirty Power at Whose Cost? A Photo Book*.

Agee, J. and W. Evans (1969), *Let Us Now Praise Famous Men: Three Tenant Families*, New York: Houghton Mifflin.

Agricola, G. (1556), *De re Metallica*.

Allen, V. L. (1981), *The Militancy of British Miners*, Shipley: The Moor Press.

Attie, S. (2008), *The Attraction of Onlookers, Aberfan: An Anatomy of a Welsh Village*, Cardigan: Parthian Press.

Augé, M. (2008), *Non Places: An Introduction to Supermodernity*, London: Verso.

Bahadur, S. (2009), *The Sound of Water*, New York: Atria.

Balsom, E. and H. Peleg (2016), 'Introduction: The Documentary Attitude', in E. Balsom and H. Peleg (eds), *Documentary Across Disciplines*, Cambridge, MA: MIT Press.

Barr, J. (1969), *Derelict Britain*, London: Penguin Books.

Barsam, R. M. (1974), *Nonfiction Film: A Critical History*, London: Allen & Unwin.

Beardmore, S. (1986), '*I Hate Green*', Part of the Valleys Project, Cardiff: Ffotogallery.

Benjamin, W. (1931), 'A Short History of Photography', in *One Way Street* (1979), London: Verso.

Bermingham, A. (1987), *Landscape and Ideology: The English Rustic Tradition, 1740–1860*, London: Thames and Hudson.

Bing, W. (2013), 'Filming a Land in Flux', *New Left Review*, 82, July–August, pp. 115–134.

Bonneuil, C. and J.-P. Fressoz (2016), *The Shock of the Anthropocene*, London: Verso.

Bryson, B. (1997), *A Walk in the Woods*, London: Penguin Books.

Bulmer, M. (1978), *Mining and Social Change*, London: Croom Helm.

Burke, E. (1757), *A Philosophical Enquiry into the Origin of Our Ideas of the Sublime and the Beautiful*, London.

Burnett, F. H. (1893), *That Lass O'Lowries: A Lancashire Story*, London: Frederick Warne & Co.

Cabuts, P. (2012), *Creative Photography and Wales: The Legacy of W. Eugene Smith in the Valleys*, Cardiff: University of Wales Press.

Clinton, H. (2017), *What Happened*, London: Simon & Schuster.

Caudill, H. M. (1962), *Night Comes to the Cumberlands: A Biography of a Depressed Area*, Canada: Little Brown.

Cohen, P. A. (2014), *History and Popular Memory: The Power of Story in Moment of Crisis*, New York: Columbia University Press.

Conn, D. (2017), 'The Scandal of Orgreave', *The Guardian*, Wednesday 17 May.

Coombes B. L. (1939), *These Poor Hands*, London: Gollancz.

Corton, C. L. (2015), *London Fog: The Biography*, Cambridge, MA: Harvard University Press.

Coyle, G. (2010), *The Riches Beneath Our Feet: How Mining Shaped Britain*, Oxford: Oxford University Press.

Cronin, A. J. (1935), *The Stars Look Down*, London: Gollancz.

Davies, A. (2006), *The Pit Brow Women of the Wigan Coalfield*, London: Tempus Publishing.

Davies, J. (2006), *The British Landscape*, London: Chris Boot.

Davies, J. (2016), *Shadow: Terrils d'Europe du Nord, Slag Heaps of Northern Europe*, Pas-de-Calais, Labanque: editions Loco.

Defoe, D. (1726), *A Tour Through the Whole Island of Great Britain*, London: Folio Edition, 1991.

Deller, J. (2002), *The English Civil War Part 11: Personal Accounts of the 1984–85 Miners' Strike*, London: Artangel.

Deller, J. (2013), *All that Is Solid Melts into Air*, London: Hayward Publishing.

Denning, M. (1996), *The Cultural Front: The Laboring of American Culture in the Twentieth Century*, London: Verso.

Dickens, C. (1854), *Hard Times*, London.

Dickens, C. (1865), *Our Mutual Friend*, London: Penguin Books, 1997.

Diehl, C. (2006), 'The Toxic Sublime', *Art in America*, Feb., pp. 118–123.

Dilnot, C. (2012), 'The Gleaners', *New Left Review*, 77, Sept./Oct., pp. 145–149.

The Economist (2014), 'Turkish Politics: Disillusioned and Divided'. 23 May.

Edensor, T. (2005), *Industrial Ruins: Space, Aesthetics and Materiality*, London: Berg.

Edwards, H. (1937), *The Good Patch*, London: Cape.

Edwards, J. A. and J. C. L. i Coit (1996), 'Mines and Quarries. Industrial Heritage Tourism', *Annals of Tourism Research*, 23(2): pp. 341–363.

Ekirch, R. (2005), *At Day's Close: A History of Nighttime*, London: Phoenix Books.

Evans, A. (2000), *Black Wound*, Exhibition Catalogue, Newport Museum and Art Gallery, Newport, Wales.

Evelyn, J. (1661), *Fumifugium, Or The Inconveniencie of the Aer and Smoak of London Dissipated*, London.

Flannery, T. (2015), *Atmosphere of Hope: Solutions to the Climate Crisis*, London: Penguin Books.

Foucault, M. (1972), *The Archeology of Knowledge*, London: Tavistock Publications.

Frank, R. (1959), *The Americans*, New York: Grove Press.

Franklin, S. (2016), *The Documentary Impulse*, London: Phaidon Press.

Frazier, La Toya R. (2017), *And from the Coaltips a Tree Will Rise*, Brusells: Musée Des Arts Contemporains de la Fédération Wallonie-Bruxelles (MAC).

Freese, B. (2003), *Coal: A Human History*, London: Heinemann.

Fuller, P. (1985), 'Black Arts: Coal and Aesthetics', in his *Images of God: The Consolations of Lost Illusions*, London: Chatto and Windus.

Gier-Viskovatoff, J. J. and A. Porter (1998), 'Women in the British Coalfields on Strike in 1926 and 19784: Documenting Lives Using Oral History and Photography', *Frontiers: A Journal of Women's Studies*, 19(2): 199–230.

Godfrey, I. (2013), *Legacy of the Mine*, South Africa: Jacana Media.

Goldblatt, D. and N. Gordimer (1973), *On the Mines*, Steidl edition 2012, Göttinggen: Steidl.

Gorman, J. (1985), *Images of Labour*, London: Scorpion Press.

Gorz, Á. (1980), *Farewell to the Working Class*, London: Pluto Press.

Grant, S. and J. Berger (1988), 'Walking Back Home', in C. Killip (eds), *In Flagrante*, London: Secker and Warburg.

Greene, G. (1972), *The Pleasure Dome: The Collected Film Criticism, 1935–40*, London: Secker & Warburg.

Gray, D. (1982), *Coal: British Mining in Art 1680–1980*, London: Arts Council of Great Britain.

Grisham, J. (2014), *Gray Mountain*, New York: Bantam Books.

Gunson, A. J and Y. Jian (2012), *Artisanal Mining in the People's Republic of China*, Report of the International Institute of Environment and Development.

Hall, D. (2012), *Working Lives: The Forgotten Voices of Britain's Post-War Working Class*, London: Bantam Press.

Hall, S. (1983/1984), 'Left in Sight', *Camerawork*, No. 29, Winter, pp. 17–19.

Halliday, S. (2001), 'Death and Miasma in Victorian London: An Obstinate Belief', *BMJ*, 22 Dec.; 323(7327): 1469–1471.

Hannavy, J. (2013), *Edwardian Mining in Old Postcards*, Wellington: PiXZ Books.

Harcup, T. (2011), 'Reporting The Voices of the Voiceless During the Miners' Strike: An Early Form of Citizen Journalism', *Journal of Media Practice*, 32(1): 27–23.

Harrison, T. (1985), *V.*, Hexam, Bloodaxe Books.

Hatakeyama, N. (2011), *Terrils*, France: Light Motiv.

Haveman, M. (2002), 'On Coal and Photographs', in Haveman, M. (ed.) *Black Smoke: Photography and Coal in the Twentieth Century*, Rotterdam: Nederlands Foto Instituut.

Hayes, B. L. (2011), *Canary in a Coal Mine*, New York: Alabaster Books.

Hewison, R. (1987), *The Heritage Industry*, London: Methuen.

HMSO (1985), Industry and the Camera, Royal Commission on the Historical Monuments of England, London.

Howard, P. (2002), 'The Eco-museum: Innovation That Risks the Future', *International Journal of Heritage Studies*, 8(1): 63–72.

Howes, C. (1989), *To Photograph Darkness*, Gloucester: Alan Sutton Press.

Huddle, R. et al. (1985), *Blood, Sweat & Tears: Photographs from the Great Miners' Strike 1984–85*, London: Artworker Books.

Hudson, D. (1972), *Munby, Man of Two Worlds: The Life and Diaries of Arthur J. Munby, 1828–1910*, London: John Murray.

Hurley, F. J. (1972), *Portrait of a Decade: Roy Stryker and the Development of Documentary Photography in the Thirties*, Baton Rouge: Louisiana State University Press.

Isherwood, C. (1939), *Goodbye to Berlin*, London: Penguin Books.

Jevons, S. (1866), *The Coal Question: An Enquiry Concerning the Progress of the Nation and the Probable Exhaustion of Our Coal Mines*, London: Macmillan.

John, A. (1984), *By the Sweat of their Brow: Women Workers at Victorian Coal Mines*, London: Routledge and Kegan Paul.

Johnson, B. (2010), '"An Upthrust into Barbarism": Coal, Trauma, and Origins of the Modern Self, 1885–1951', *Journal of American Culture*, 33(4) (Dec.): 265–279.

Jones, M. H. (1925), *Autobiography of Mother Jones*, Chicago: Charles Kerr.

Jones, Rev. H. (1893), *Illustrated London News*, 18 November.

Keating, J. (1900), *Son of Judith: A Tale of the Welsh Mining Valleys*, London: Simpkin Marshall.

Keating, J. (1916), *My Struggle for Life*, London: Simpkin Marshall.

Killip, C. (2011), *Seacoal*, Göttingen: Steidl Publishers.

Klein, M. (2013), 'Of Politics and Poetry: The Dilemma of the Photo League', in M. Klein and C. Evans (eds), *The Radical Camera: New York's Photo League, 1936–1951*, New Haven: Yale University Press.

Klingender, F. D. (1947), *Art and the Industrial Revolution*, Revised edition 1972, London: Paladin Press.

Koudelka, J. (1994), *The Black Triangle, The Foothills of the Ore Mountains*, Prague: Sagga Co-operative.

Koudelka, J. (1998), *Reconnaissance Wales*, Cardiff: Ffotogallery.

Kranitz, S. (2014), *Oxford American* Online Photography Feature, 'Eyes on the South', March, oxfordamerican.org.

Lawrence, D. H. (1929), 'Mining and the Nottinghamshire Countryside', *Phoenix: The Posthumous Papers of D.H. Lawrence*, 1936, London: Heinemann.

Leeworthy, D. (2009), 'Characterising South Wales and Cape Breton as an Industrial Frontier', *LLafur*, 10(2): 53–71.

Light, K. and M. Light (2006), *Coal Hollow: Photographs and Oral Histories*, Los Angeles: University of California Press.

Llewellyn, R. (1939), *How Green Was My Valley*, London: Michael Joseph.

Llewellyn, R. (1975), *Green, Green My Valley Now*, London: Michael Joseph.

Lockwood, A. H. (2012), *The Silent Epidemic: Coal and the Hidden Threat to Health*, Cambridge, MA: MIT Press.

Long, P. (1989), *Where the Sun Never Shines: A History of America's Bloody Coal Industry*, New York: Paragon House.

Lowndes, M. B. (1914), *The Lodger*, New York: Scribner.

MacCannell, D. (1976), *The Tourist: A New Theory of the Leisure Class*, London: Macmillan.

Madgwick, G. (2016), *Aberfan: A Story of Survival, Love and Community in One of Britain's Worst Disasters*, Aberystwyth: Y Lolfa.

Manifesta 9 (2012), *The Deep of the Modern: A Subcyclopaedia*, Belgium: Ghenk.

Marchand, Y. and R. Meffre (2013), *Gunkanjima*, Göttingen: Steidl.

Margolis. M. (1998), 'Picturing Labor, A Visual Ethnography of the Coal Mine Labor Process', *Visual Sociology*, 13(2): 5–32.

Martin, R. (2015), *Coal Wars: The Future of Energy and the Fate of the Planet*, London: St. Martin's Press.

Massey, P. (1937), *Portrait of a Mining Town*, London: Fact.

McLean, I. and M. Johnes (2000), *Aberfan: Government and Disaster*, Cardiff: Welsh Academic Press.

A Medical Survey of the Bituminous Coal Industry (1947), Coal Mines Administration (Washington, D.C.: Department of the Interior).

Miller, J. (1986), *You Can't Kill the Spirit*, London: Women's Press.

Monbiot, G. (2017), *How Did We Get Into This Mess?*, London: Verso.

Morris, H. L. (1934), *The Plight of the Bituminous Coal Miner*, Philadelphia: University of Pennsylvania Press.

Mosely, S. (2004), *Public Perceptions of Smoke Pollution in Victorian Manchester*, in Melanie DuPuis (ed.), *Smoke and Mirrors: The Politics and Culture of Air Pollution*, New York: New York University Press.

Myers, D., E. M. Walker and N. Vollmer (2017), *Carrying Coal to Columbus: Mining in the Hocking Valley*, Stroud: The History Press.

Nield, T. (2014), *Underlands: A Journey Through Britain's Lost Landscapes*, London: Granta.

Nye, D. E. (1994), *American Technological Sublime*: Cambridge, MA: MIT Press.

O'Boyle, S. (2010), *Modern Ruins: Portraits of Place in the Mid Atlantic Region*, Pennsylvania University Press, Keystone Books.

Ohlin, A. (2002), 'Andreas Gursky and the Contemporary Sublime', *Art Journal*, 61(4), Winter: 22–35.

Oldham, C. (2015), *In Loving Memory of Work: A Visual Record of the UK Miners' Strike 1984–1985*, London: Verso.

Oldham, C. (2016), 'Elbow Grease: Perseverance and Proper Graft', *Ascendance*, Issue 1.

Orwell, G. (1937), *The Road to Wigan Pier*, London: Victor Gollancz.

Osnos, E. (2011), *The New Yorker*, 1 November.

O'Sullivan, J. (2001), *Photographic History of Mining in South Wales*, Stroud: Sutton Publishing.

Parr, M. and L. Wassink (2015), *The Chinese Photobook from the 1900s to the Present*, New York: Aperture.

Pattison, K. (2010), *No Redemption*, London: Flambard Books.

Pauli, L. (2003), *Manufactured Landscapes: The Photographs of Edward Burtynsky*, Ottawa: National Gallery of Canada.

Pearse, G., D. McKnight and B. Burton (2013), *Big Coal: Australia's Dirtiest Habit*, Sydney: New South Press.

Pollock, V. L. (2009), 'Dislocated Narratives and Sites of Memory: Amateur Photographic Surveys in Britain 1889–1897', *Visual Culture in Britain*. Vol. 10:1, 1–26.

Power, R. (2008), '"After the Black Gold": A View of Mining Heritage from Coalfield Areas in Britain', *Folklore*, 119(2): 160–181.

Price, D. (1986), 'How Green Was My Valley: A Romance of Wales', in Radford J. (ed.), *The Progress of Romance: The Politics of Popular Fiction*, London: Routledge.

Priestley, J. B. (1934), *English Journey*, London: Heinemann.

Radetsky, P. (2007), *The Soviet Image: A Hundred Years of Photographs from Inside the TASS Archives*, San Francisco: Chronicle Books.

Rae, P. (1999), 'Orwell's Heart of Darkness: The Road to Wigan Pier as Modernist Anthropology', *Prose Studies*, 22(1): 71–102.

Reeves, C. (n.d.), 'Redressing the Balance: Making the Miners Campaign Tapes,' in a booklet accompanying the tapes, distributed by the British Film Institute.

Report on Child Labour (1842), The Condition and Treatment of the Children Employed in the Mines and Colliers of the United Kingdom, London: William Strange and Co.

Rose, J. (2002), '*The Intellectual Life of the British Working Classes*', New Haven: Yale University Press.

Roth, M. S. (1992), 'The Time of Nostalgia, Medicine, History and Normality in 19th-century France', *Time and Society*, 1(2), May.

Rothstein, A. (1986), *Documentary Photography*, Boston: Focal Press.

Samuel, R. (1994), *Theatres of Memory*, London: Verso.

Schuster, J. (2013), 'Between Manufacturing and Landscapes: Edward Burtynsky and the Photography of Ecology', *Photography and Culture*, 6(2), July.

Scott, C. (2009), 'Contemporary Expressions of Coal Mining Heritage in the Durham Coalfield: The Creation of New Identities', *Folk Life, Journal of Ethnological Studies*, 47(1): 66–75.

Scott, R. R. (2010), *Removing Mountains: Extracting Nature and Identity in the Appalachian Coalfields*, Minneapolis: University of Minnesota Press.

Sekula, A. (1983), 'Photography between labour and capital', in B. H. D. Buchloh and R. Wilkie (eds), *Mining Photographs and Other Pictures 1948–1968: A Selection from the Negative Archives of Shedden Studio Glace Bay Cape Breton*, Nova Scotia: The Press of Nova Scotia College of Art and Design and The University of Cape Breton Press.

Sekula, A. (1986), 'Geography Lesson: Canadian Notes', *Ten.8: International Photographer Magazine*, No. 29. 1988.

Shogan, R. (2004), *The Battle of Blair Mountain: The Story of America's Largest Labor Uprising*, New York: Basic Books.

Sinclair, U. (1917), *King Coal*, Northridge, CA: AEgypan Press.

Spender, H. (1978), 'Humphrey Spender: MO Photographer, *Camerawork*, 11, pp. 6–7.

Steichen, E. (1938), 'The FSA Photographers', *U.S. Camera Magazine*, New York: William Morrow and Co.

Stewart, S. (1993), *On Longing: Narratives of the Miniature, the Gigantic, the Souvenir, the Collection*, Durham and London: Duke University Press.

Stimson, B. (2006), *The Pivot of the World: Photography and Its Nation*, Cambridge, MA: MIT Press.

Stradling, D. and P. Thorsheim (1999), 'The Smoke of Great Cities: British and American Efforts to Control Air Pollution 1860–1914', *Environmental History*, 4(1) (Jan.): 6–31.

Tabuchi, H. (2017), 'As Beijing Joins Climate Fight, Chinese Companies Build Coal Plants', *New York Times*, 1 July.

Tallichet, S. (2006), *Daughters of the Mountain: Women Coal Miners in Central Appalachia*, Pennsylvania: Pennsylvania State University Press.

Thompson, E. P. (1963), *The Making of the English Working Class*, London: Gollancz.

Thompson, P. (1988), *The Voice of the Past: Oral History*, Oxford: OUP.

Thomson, J. (1877), *Street Life in London*: Low, Marston, Searle and Rivington.

Thorsheim, P. (2006), *Inventing Pollution: Coal, Smoke and Culture in Britain Since 1800*, Athens, OH: Ohio University Press.

Trachtenberg, A. (1977), 'Ever- The Human Document', in *America and Lewis Hine: Photographs 1904–1940*, New York: Aperture Press.

Tudor-Hart, E. (2013), *In the Shadow of Tyranny*, Ostfildern: Hatje Cantz.

Vance, J. D. (2016), *Hillbilly Elegy: A Memoir of a Family and a Culture in Crisis*, London: William Collins.

Vasilyeva, W. (2012), Introduction to, *Soviet Era by Markiv-Grinberg*, Moscow: Lumiere Brothers Center for Photography.

Walker, I. (2007), *So Exotic, So Homemade: Surrealism, Englishness and Documentary Photography*, Manchester: Manchester University Press.

War on Want (2016), *The New Colonialism: Britain's Scramble for Africa's Energy and Mineral Resources*, London: War on Want.

Wells, L. (2011), *Land Matters: Landscape Photography, Culture and Identity*, London: I.B. Taurus.

Williams, J. (1980), 'The Coal Industry, 1750–1914', *Glamorgan County History*, 5, pp. 147–159.

Wolf, E. (2011), 'The Soviet Union: From Worker to Proletarian Photography', Soviet Photo Correspondents, Academia.edu.

Wright, M. (2014), *Coal: The Rise and Fall of King Coal in New Zealand*, Auckland: Bateman Ltd.

Xinyu, Lu (2005), 'Ruins of the Future: Class and History in Wang Bing's Tiexi District', *New Left Review*, 31, January–February, pp. 125–136.

Zheng, L. (2004), *The Chinese*, Göttingen: Steidl.

Zola, É. (1885), *Germinal*, London: Penguin Books, 1955.

Zukin, S. (1991), *Landscapes of Power: From Detroit to Disney World*, Berkley: University of California Press.

Zweig, F. (1948), *Men in the Pits*, London: Gollancz.

FILMS

Title, Date, Director.
After Coal: Welsh and Appalachian Mining Communities (2016) Tom Hansell.
The Bay Boy (1984) Daniel Petrie.
Behemoth (2015) Zhao Liang.
Black Hole: Transforming a Forest into a Coalmine (2015) João Dujon Pereira.
Centralia: Pennsylvania's Lost Town (2017) Joseph Sapienza.
Coal Face (1935) Alberto Cavalcanti.
Coal Miner's Daughter (1980) Michael Apted.
Coal Money (2009) Wang Bing.
Guilty Chimneys (1954) Gerard Bryant.
Hard Coal: The Last of the Bootleg Miners (2008) Marc Brodzik.
Harlan County, USA (1976) Barbara Kopple.
Kameradschaft (1931) G. W. Pabst.
The Lodger (1927) Alfred Hitchcock.
The Lodger (2009) David Ondaatje.
Matewan (1987) John Sales.
Misère au Borinage (1933) Joris Ivens and Henri Storck.
Pride (2014) Matthew Warchus.
The Proud Valley (1940) Pen Tennyson.
Pyramiden (2016) David Beazley.
Seacoal-UK (2006) Chloë Mercier and Edward King.
The Stars Look Down (1940) Carol Reed.
Tie Xi Qu: West of the Tracks (2002) Wang Bing.
Which Side Are You On? (1985) Ken Loach.

INDEX